Beginner's Guide to **CNC**
MACHINING
in Wood

T0333347

Beginner's Guide to CNC Machining in Wood

Understanding the Machines, Tools, and Software, Plus Projects to Make

FOX CHAPEL
PUBLISHING

Ralph Bagnall

Beginner's Guide to CNC Machining in Wood is an original work, first published in 2022 by Fox Chapel Publishing Company, Inc. The patterns contained herein are copyrighted by the author. Readers may make copies of these patterns for personal use. The patterns themselves, however, are not to be duplicated for resale or distribution under any circumstances. Any such copying is a violation of copyright law.

Unless otherwise noted, all photography by Ralph Bagnall. The following images are credited to Shutterstock.com and their respective creators: page 7 top (sawdust): PurMoon; page 23: Tomasz Nieweglowski

ISBN 978-1-4971-0058-9

Note: The imperial measurements in this book have been converted to metric measurements as accurately as possible. It must be noted that the imperial measurements for T-bolts do not have a direct metric conversion. A close substitute has been provided.

Library of Congress Control Number: 2021931484

To learn more about the other great books from Fox Chapel Publishing, or to find a retailer near you, call toll-free 800-457-9112 or visit us at *www.FoxChapelPublishing.com*.

We are always looking for talented authors. To submit an idea, please send a brief inquiry to acquisitions@foxchapelpublishing.com.

Printed in the United States of America
First printing

CONTENTS

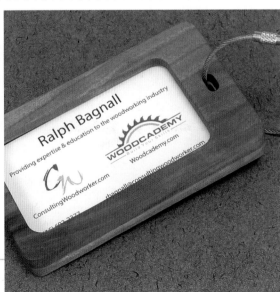

Introduction

I have a brother who is a machinist, so I've known about computer-controlled milling machines for decades. I first became aware of computerized milling machines for woodworking in the mid-1990s. The company I was working for at the time had purchased a CNC (Computer Numeric Control) machine for milling cabinet parts from sheet goods, and I quickly volunteered to be one of the employees trained to operate the system. I soon became the prime operator and, before long, the only one.

Over the years, I have learned to install, service, and support these machines and have trained many people around the United States and Caribbean to program and operate CNCs in a variety of woodworking companies. Yet, as much as I wanted to, owning a CNC of my own was not really practical for many years due to both cost and size.

As with many forms of technology, however, CNC machines were becoming less expensive and being made smaller each year. Finally, in 2010, I acquired my CNC Shark from Rockler®. I went on to work with Rockler to create a series of instructional videos for new users. I also regularly contribute programs used by Rockler's *Woodworker's Journal Magazine* for its online videos featuring CNC projects.

Both the machines and the software have continued to improve dramatically over the years, making those original videos more and more obsolete. So, when Fox Chapel Publishing approached me about writing a new manual for the beginning CNC user, I jumped at the chance.

This book was created to help you get started understanding and using your CNC machine. I have been programming and running CNCs for more than twenty years, and I still am learning new techniques and developing more diverse and interesting projects.

The machines and the software that programs them are constantly evolving and doing more, widening the abilities of even beginners very quickly.

Use this book in the spirit intended: as a practical guide to help you flatten the learning curve as you grow into your machine. I had a lot of fun creating this book and hope it helps you enjoy the process of learning. This book should be viewed as a launch pad to give you the tools you need to control your milling and inspire you to use them in ways no one else may have done yet.

Ralph Bagnall

Here I am with my CNC machine.

About the Author

Ralph Bagnall has been a professional woodworker for 30 years and began working with CNC machines around 1996. He has been teaching and consulting to the trade since 2007 with an emphasis on CNC operations and management. He began writing woodworking articles in 2000 and has been a speaker at many events, from the International Woodworking Fair to The Virtual Wood Show; he began hosting the Woodcademy TV show in 2017. Ralph and his wife currently reside in central Florida.

Gallery of CNC Projects

Most people buy their first CNC with a specific purpose in mind, but once you learn how your machine works, you will come up with many more. I've gathered this gallery of CNC-made projects to show you just a few of the many possibilities a CNC machine offers. I've included some decorative projects, jigs and fixtures, game boards and furniture pieces. There's even a tin canister lid that was embossed on a CNC-milled form. Your imagination really can run wild once you begin to master programming and operating your CNC machine.

This tissue box cover takes advantage of the texturing feature in the CAD/CAM software to add interest to the panel spaces.

A pair of small oven sticks for a toaster oven are easy to duplicate exactly with a CNC machine.

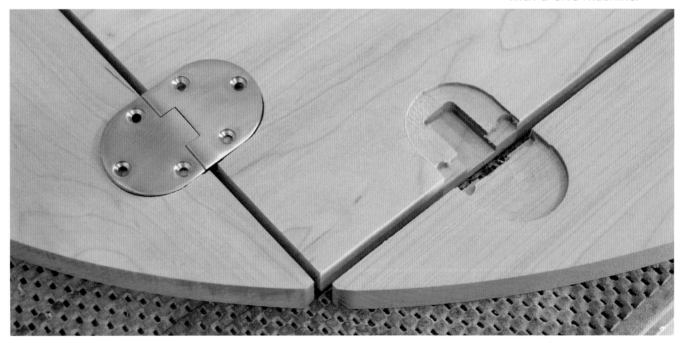

The CNC excels at accurately cutting the complex mortises required by this butler tray's special hinge.

A CNC machine equipped with V-bits can carve images with very fine details.

PHOTO CREDIT: SHOPBOT TOOLS

My popular blade storage box has pullout blade holders with staggered tabs for labels. The computer and CNC ensure that all of these components fit properly.

This chess piece shows how a CNC allows you to use short offcuts that may otherwise be thrown away.

The sides and ends of this basket feature a woven texture that you may not be able to create without a CNC machine.

A CNC can help you make fixtures that improve other woodworking tools. This shop-made steady rest keeps stock from vibrating during turning on a lathe.

With the right tools, a CNC can replicate a wide range of moldings and appliques you can use with all sorts of projects.

This pretty mosaic looks like the work of a master mason, but it's the product of careful CNC programming.

Two pieces of stock embellished with numerous carving techniques produces a beautiful clock.

Before cutting any expensive lumber, I used my CNC to fabricate this scale-model deck chair to ensure all the parts would fold and unfold properly.

Your CNC and a little imagination can transform an ordinary cut off into a prized possession,

CNC-cut pockets work very well with cast resin for creating many decorative items.

The gears on this compass actually work to lock the legs at specific measurements. All the parts were cut on a CNC machine.

This tea tray features curved handles, pierced rails and a carved base, all made on the CNC machine.

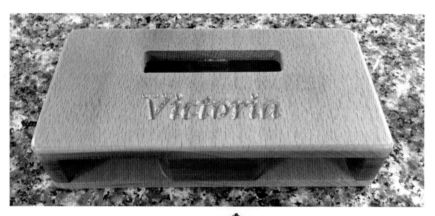

A personalized passive amplifier for a cell phone demonstrates a CNC's custom lettering capabilities.

PHOTO CREDIT: SHOPBOT TOOLS

A simple serving tray becomes a unique gift when customized with a CNC.

It's hard to imagine that this dramatic carving can be accomplished with a single router bit.

Your CNC can cut and engrave non-ferrous metals with the proper tooling.

Game boards can be shaped, lettered, engraved and scooped out easily on your CNC machine.

Gallery of CNC Projects

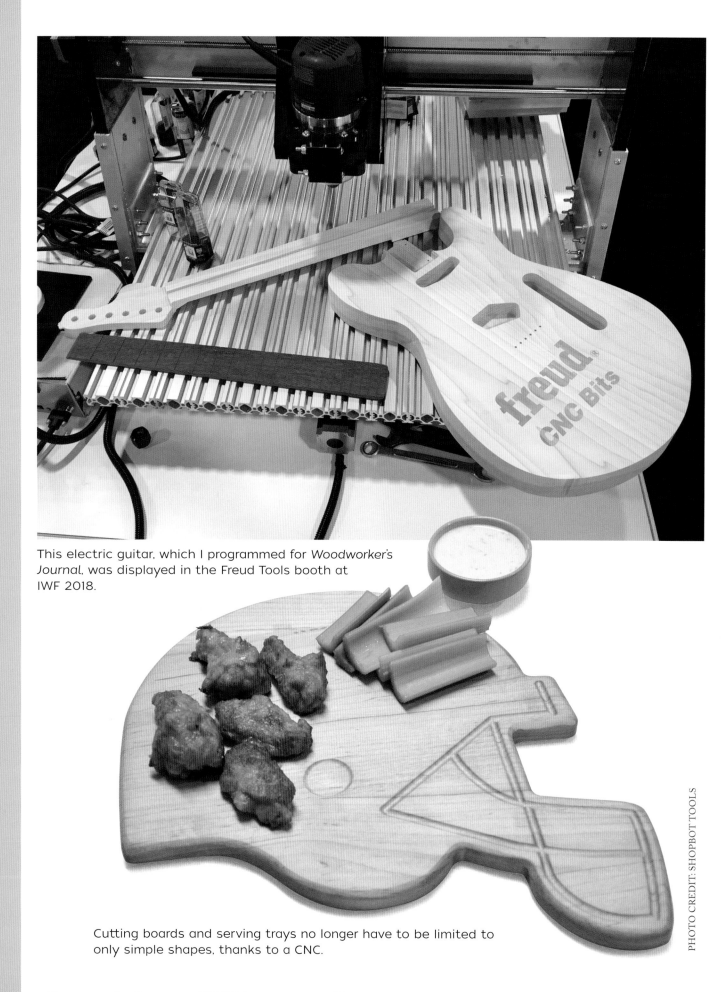

This electric guitar, which I programmed for *Woodworker's Journal*, was displayed in the Freud Tools booth at IWF 2018.

Cutting boards and serving trays no longer have to be limited to only simple shapes, thanks to a CNC.

Your CNC can be used to make jigs and templates that are then used to locate and hold parts for milling.

Creating fun and educational puzzles for kids has never been easier, thanks to a CNC's precision cutting capabilities.

How to Use This Book

As I have trained CNC operators over the years, I have found that it is very easy for students to get lost in the unique jargon and myriad of technical details needed to become successful. Experience has shown that teaching techniques inside a specific project makes it easier to both understand and retain the information presented. Think back to grade school geometry: the formula for finding the area of an object means nothing by itself, but when applied to finding out how much paint is needed for a tree house, understanding is much easier. In that light, this book will use practical projects to teach specific skills, and we will not separate the drawing from the toolpathing from the physical running of the machine. All three are tied very closely together, so they tend to be best presented together within a project. Once you've mastered the theory by working through the projects in this book, you can go on to apply the techniques to your own projects, further strengthening your CNC operating skills.

Did You Know?

The acronym **CNC** stands for "computer numerical control." A CNC machinist uses various computer software programs to design objects or create patterns on a block of material (we'll be using wood for the purposes of this book) that dictates how the machine and tools will move to create a project.

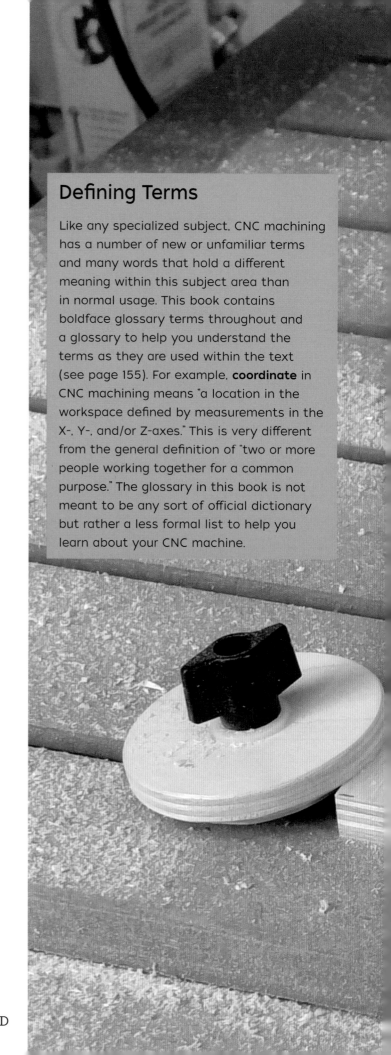

Defining Terms

Like any specialized subject, CNC machining has a number of new or unfamiliar terms and many words that hold a different meaning within this subject area than in normal usage. This book contains boldface glossary terms throughout and a glossary to help you understand the terms as they are used within the text (see page 155). For example, **coordinate** in CNC machining means "a location in the workspace defined by measurements in the X-, Y-, and/or Z-axes." This is very different from the general definition of "two or more people working together for a common purpose." The glossary in this book is not meant to be any sort of official dictionary but rather a less formal list to help you learn about your CNC machine.

Frequently Asked Questions

Over the years I've been selling CNC systems to businesses as well as training professionals and hobbyists in their use, I have been asked the same questions that you are probably asking as you consider buying your first CNC. Let's start this book by answering them.

Why do I need CNC machine?

This piecrust tabletop was flattened, dished out, carved on both sides and cut out on my CNC machine.

Think of your CNC as an almost unlimited jig. It can be "adjusted" to fit many different needs. Any task it can do once it can accurately repeat over and over. No matter how many programs you write for it, they take up no additional shop space, whereas jigs will just keep accumulating on shelves and in cabinets. You can certainly use the CNC to make many of the actual wood parts you need for your projects, like the pie crust table top shown above. Several of the projects in this book are made almost entirely on the CNC machine with only finishing steps needed after milling.

But in truth, my CNC gets used a lot more to make items that improve the more traditional woodworking that I enjoy doing. I no longer need to hand-make throat plate inserts for some of my woodworking machines. Once programmed into your computer, you can quickly make a new band saw or table saw throat plate any time yours gets worn, and it will always fit properly. Later in this book we will work through programming and milling a Zero Clearance Insert for your table saw (see page 78).

I do a lot of template routing in my shop, and my CNC makes it easy to create the templates quickly and accurately. Working your way through this book you will make a push stick, hold downs, a bit guard, and other shop projects. My experience with pros and hobbyists is that they buy a CNC with a particular task in mind but quickly find many more uses once they get started.

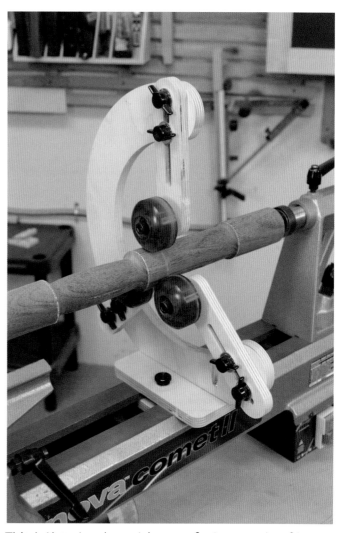

This lathe steady rest is a perfect example of how a CNC machine can improve the functionality of other more traditional woodworking machines. The lathe fixture would have been very difficult to make accurately without a CNC.

How much computer and general woodworking machine experience should I have to get started with a CNC machine?

I've often been asked by business owners, "Do I need to hire a computer programmer to run the machine?" The answer is no. The CNC is a router. It uses router bits to cut wood and plastics, same as you would with your handheld router. It is far easier to teach a woodworker to program the CNC than it is to teach a computer programmer how to work in wood. This was true 20 years ago when I got started and it is even more true now because the CAD/CAM software continues to become easier to learn and use. How a router bit cuts, how fast or slow to move it, what speed to set it at, issues of cutting with or against the wood grain — all of these factors are more difficult to teach someone who is not a woodworker.

Can I make money with my CNC?

Your CNC can make you money just as easily as your table saw, band saw, or any other tool in your shop. Many people now make all sorts of products using a CNC and sell them on the craft circuit, through Etsy or their own websites. I know of several professional shops with CNCs that do nothing but cut parts for other woodshops. I make some money writing programs for other people to run on their machines as well as through articles and books like this one. It is certainly possible to make money with a CNC machine. How much you can make, or whether you even want to, is a much more difficult question that only you can really answer.

This carved tissue box cover is a CNC project that makes a great gift or item to sell. The CAD/CAM programming allows for a basic design that can be easily customized with names, logos or sayings carved into the panels instead of texturing.

Metric Conversions

Throughout this book dimensions are being presented in common fractions and Imperial decimal numbers. Metric conversions of these numbers follow each where appropriate.

In many cases these will be direct conversions; ¼" becomes 6.35mm. But when discussing router bits, there is no 6.35mm bit made. The common metric equivalent would be a 6mm or possibly 8mm router bit, so I will use these numbers as appropriate.

Finally, when discussing workpiece blank sizes, a 10" blank can be displayed as 250mm rather than 254mm, since that is what the common metric measurement would be when using a tape measure or setting a rip fence. The parts will still fit inside common blank measurement.

CHAPTER 1

GETTING STARTED

There are many things you need to consider as you prepare to buy and begin using your first CNC machine. What you want to make, your workspace, your experience and budget should be taken into account, but at this point you may not even know what questions to ask. This first chapter will walk you through a series of selections to help you decide which type and size of machine best suits your needs. There is a lot to learn about programs, CNC operations, bit selection and shop setup, so this chapter will also provide an overview of the process behind programming and running a CNC machine. Reading and understanding this chapter before you buy a machine will help you make the most appropriate choice for your shop.

Setting Up a Workspace

When setting up a CNC machine, you need to consider not only the space, lighting and power required but also the dust collection and noise issues that will arise. Expect that your CNC machine will create about the same noise and mess as a handheld router. The noise and dust issues both need to be addressed, especially in an attached garage or basement shop in your home.

Collecting Dust

With very few exceptions, no CNC machine at any price point has excellent dust collection. They can generally manage to keep most of the dust out of the air, but do not expect a clean surface when the machine finishes milling a program. Many early benchtop machines did not even try to collect dust, although most machines on the market today include some type of dust hood.

The main problem with CNC dust collection efficiency is airflow. The typical setup is a 2" (50.8mm) or even a 4" (101.6mm) hose leading into the work area near the bit, but it is trying to collect dust from a large open area. There is simply too much open airspace for the hose to effectively collect all the dust being generated. The dust hood may be able to capture the lightest particles, but not much more. Enclosing the space around the bit helps with this problem. The adapter that I built for my CNC Shark is commonly called a **dust shoe**. Dust shoes are designed to cordon off the immediate work area so the air contained within the shoe moves more quickly and carries larger dust particles. But dust shoes are not perfect. The vacuum hose still has to remove air from the entire volume underneath the dust shoe, and it is just too big a change in air volume for it to maintain proper suction. The shoe has to be large enough for both the router motor and the vacuum hose, and this area will be at least twice the size of the hose, whatever hose diameter is used.

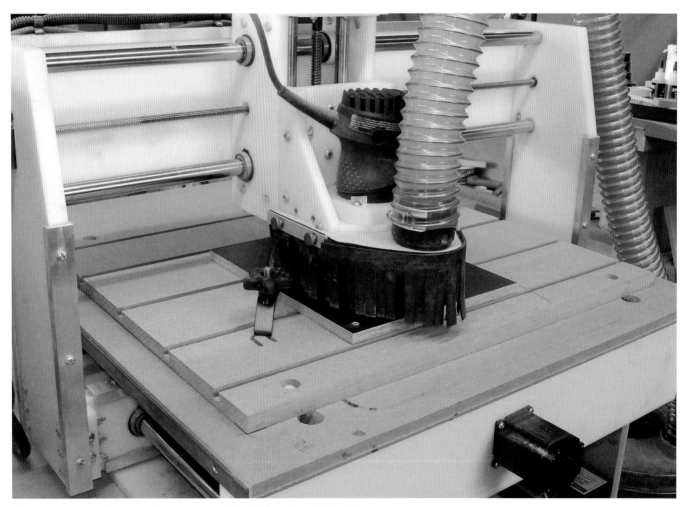

Shown here is the dust shoe that I built for my CNC Shark.

Dust hoods all require some type of flexible border to contain chips and limit air loss. These borders form a flexible connection between the top of the dust shoe and the workpiece. The better the connection, the more dust is removed by the vacuum. Mine uses foam rubber strips, while other machines might be equipped with vinyl or brush-style bristles. These borders help contain the flying chips and minimize the volume of airflow needed. The dust shoe is typically mounted to the bracket holding the router motor or even the motor itself. The distance between the dust shoe and the top of the workpiece changes with the length of the bit being used and with the depth of cut, so brushes are rarely at the optimal position to trap dust. And when the bit is cutting near the edges of a part, much of the dust shoe will be over open space, defeating the brushes. Effective dust collection at the bit is a problem with all CNC machines, even at the industrial level.

A full enclosure that covers the machine may be the best option for mitigating dust collection issues, especially if your shop is in your home. Your dust collector can be connected just to the dust shoe on your machine, or if your dust collector has the capacity, you can also have a port to evacuate the entire enclosure. Many manufacturers offer boxes specifically designed for their machines. They are not inexpensive, but containing the mess may be well worth the cost. Many CNC users also build their own enclosures. It is well designed with excellent windows to monitor the machine as it runs, and even includes controls built into the enclosure. Your workspace situation and personal preferences will determine which dust control solution is best for you, should you opt to use a full enclosure.

Noise Control

Noise will also need to be dealt with, whether your CNC is located in a home shop or even in a separate space. Many woodworking machines operate at ear-damaging noise levels, and remember that your CNC will be at least as loud as a handheld router. The CDC states that noises over 70 decibels will damage hearing over

PHOTO COURTESY OF SHOP HACKS™ (*WWW.SHOPHACKS.COM*)

A CNC machine enclosure made to contain both dust and noise.

time. The real noise you hear when routing is not from the router motor or spindle, but from the bit cutting the wood. You should, of course, be wearing hearing protection in the shop when any machines are running, but noise carries. Family members and neighbors may not be happy with your CNC producing loud noises for hours on end.

Working with doors closed in a well-insulated room will help keep neighbors from complaining, but if you want a more comfortable work area when the CNC is running, a full enclosure is about the only way to effectively control the noise. This sort of enclosure may be an aluminum frame with clear polycarbonate plastic panels, or a plywood box with windows in it. Your enclosure does not need to be elaborate, but it should have plenty of windows to be able to monitor the CNC as it runs. It will need a door that allows parts to be loaded and unloaded easily, bits to be changed and so forth. It should also be reasonably easy to remove the enclosure for cleaning and maintaining your CNC machine.

Take the time to carefully consider your work space situation as you shop for your first CNC. Woodworking with power equipment is usually noisy and messy, and your CNC will be no exception. So planning for dust and noise control before you buy will help you choose your best options.

Setting Up a Workspace 25

Choosing Your First Machine

Choosing your first CNC system can be intimidating. There are many options to choose from, and a poor choice can be costly. We will work through the major options to narrow your choices. It is not the purpose of this book to recommend any brand or model of CNC, or for that matter tooling, software or accessories. I want to provide you with as much information as I can so that you can make the best choices for your shop. I will, however, limit this discussion to what are commonly called "benchtop" CNC models. These tend to be machines with no more than 48 inches (1.2 meters) of working area in any direction. There are a few exceptions to this size limit (see pages 33-35), but those also are still home shop models. Fortunately, there are plenty of choices for you within this group. You can spend as little or as much as you wish depending on size, features and whether you want to buy a ready-made CNC system or build a customized system from components.

Laguna Tools makes full-size industrial CNC machines as well as an iQ series benchtop model.

What Will You Make?

What's important to keep in mind as you shop is what you plan to use the CNC for. The sort of machine you would use for carving signs will be different from type best suited to mill furniture parts. You will find many other uses for your new CNC as you become experienced with it, but keeping your main goals in mind will be most helpful to guide your choices as you work through the rest of these questions. Milling table legs will mean your stock may be 2" (50.8mm) or thicker, so this would require a machine with more Z-axis travel (cutting depth capacity) than many machines offer. Carving programs that would be used for making signs tend to take hours to run, so if that is your ambition, you should consider buying a machine that uses a spindle to drive the bit. A proportionally smaller CNC will be well suited for making templates while taking less space than a bigger machine. Working area and price are often the primary features advertised for CNC machines, but looking specifically at the details that matter most to you will yield a better fit for your shop.

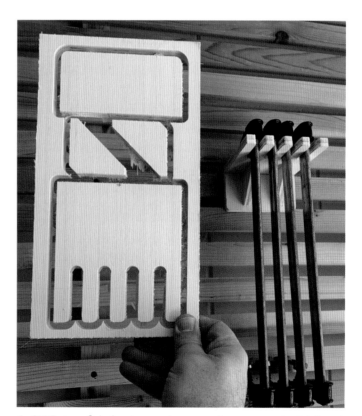

A CNC can fabricate custom parts as needed. I tend to use my CNC extensively to make jigs, fixtures, templates and storage solutions for woodworking.

Build or Buy?

If you're mechanically inclined and enjoy building projects from kits, you could consider building your own CNC instead of opting for a system that comes assembled. A kit could provide a larger, better equipped CNC within your budget, and building it yourself will teach you a lot about how it works. This will be helpful when your machine needs maintenance or repair. But keep in mind that while components from a kit manufacturer may be covered by a warranty, your skill in assembling the kit will not be warranted, and mistakes could be costly. Most of the kit manufacturers I have talked to are enthusiasts themselves and are very helpful in answering questions. And there are usually online forums where customers share tips and information.

Buying a preassembled machine is the more typical option. These are "almost ready to run". Like buying a table saw, it may need some assembly out of the box, but the factory will have done most of the assembly for you. I used to install and service industrial CNC machines, so I certainly can build from a kit, but I still chose a turn-key model for my first CNC because I wanted to spend my time using it instead of building it. If you prefer to spend your shop time making projects or just don't want to build your CNC before you can use it, then buy an almost-ready-to run-system.

PHOTO COURTESY OF GATTON CNC AND TREVOR CARTER

Building your own CNC from a kit is a viable option. Trevor Carter constructed this machine from a Gatton CNC kit that is as beautiful as it is functional.

Budget

Just as with any woodworking equipment, the cost of a CNC system can vary quite a lot depending on size and features. The open-source Maslow kit system (see page 35) starts at under $500 USD, but this figure does not include the router, bits, materials to build the machine's body and many other expenses you will see before cutting the first part. Dave Gatton of Gatton CNC says that his average CNC kit costs between $1,200 and $2,000 USD. On the other hand, a preassembled 24" (609.6mm) by 36" (914.4mm) ShopBot with an automatic tool changer will come with most everything needed and cost close to $15,000 USD, but that's on the high end of what you could spend. More practically speaking, a ready-to-run system will average from around $1,500 for a NextWave Shark SD100 with a 12" (304.8mm) by 13" (330.2mm) work zone to about $7,000 for a well-equipped Axiom 24" (609.6mm) by 36" (914.4mm) system.

PHOTO CREDIT: NEXT WAVE AUTOMATION

NextWave Automation's SD100 is a fully capable, pre-assembled CNC that sells for around $1,500 USD.

What Size?

Another consideration to make is the physical size of the CNC machine as well as what its work capacity is. Buying the most capacity you can afford may seem like a good idea, but you likely will have to make compromises in your shop space to fit a machine that large. You can buy a CNC that can be stored in a cabinet when not needed or get one with a footprint as large as a cabinet saw. Knowing what you want to make with your CNC will help guide you here. As you are shopping machines, take note of the "working" capacity. This figure may be listed as "usable area", "work zone", "working envelope" or something similar, but I am referring to the size of the actual area that the router bit can cut within. The overall footprint of a CNC is not the same as the size of the part it can actually process.

With a working area of 24" x 18" (609.6mm x 457.2mm) the Shopbot Desktop CNC offers good mid-size capacity for a benchtop CNC.

CAD/CAM Software

Most ready-to-use, consumer-grade CNC systems will come with a CAD/CAM program included or available as an option. The kit manufacturers generally offer CAD/CAM options as well. The good news is that you do not really need to worry about which CAD/CAM packages will work with the machine you buy. Most CAD/CAM systems work with the majority of CNC machines. Two notable exceptions are the Shaper Origin (see page 34) and Carvewright (see page 35) machines. Both of these can accept CAD designs and/or 3D models from outside sources, but the actual CAM programming of the toolpaths is done in a proprietary software unique to the brands. This isn't a point of concern, however. Shaper Origin and Carvewright are well-established companies, provide ample tech support for consumer users and have active online user groups that are remarkably helpful, especially for new users.

Ironically, as the number of machine brands and types continues to grow, the options for CAD/CAM software packages to run them is shrinking. As of this writing, there are really three major brands to choose from: BobCAD, Fusion 360 and various Vectric products such as VCarve Pro.

BobCAD has evolved over the years from a free utility to a powerful system that can control even highly complex industrial CNC machines. BobCAD Mill is the company's entry-level program designed for use with benchtop CNCs. BobCAD is now subscription-based and costs about $250 USD per year. Fusion 360 is offered by AutoDesk, the company that makes AutoCAD. It's also a subscription-based software. The basic subscription is $495 USD per year, but for personal and non-commercial use it is free, which is a helpful if you are just using your CNC to augment your hobby woodworking. The subscription model of these two CAD/CAM programs may put off some buyers who prefer to own their software, but keep in mind that a subscription means you will always have the most up-to-date version of the software without needing to upgrade every year or two.

BobCAD is a well-established CAD/CAM software brand that's suitable for hobby use.

AutoDesk provides a free version of its Fusion 360 CAD/CAM software for personal use.

Both BobCAD and Fusion 360 use 3D modeling as the basis for their CAD programming. This can be an advantage when using them for complex 3D carvings or if you want to program other machines like a 3D printer. But it does mean a longer learning curve for the beginner over simpler 2D CAD drawing, which I've found to be sufficient for the vast majority of my CNC programs over the years. Even though I have access to very expensive 3D modeling software, it's simply faster to work in 2D unless there is a reason that 3D is required. 3D modeling programs also require a more powerful computer to run them than 2D CAD systems.

Over the last few years, VCarve Pro has become the default CAD/CAM software package for commercial CNC systems. Even expensive industrial CNCs often come with VCarve Pro, and it is the software I use throughout this book. I have created programs for many different CNC machines with it and have yet to find an incompatible brand. The CAD drawing in VCarve Pro is 2D, but VCarve is exceptional at creating tool paths for 3D models that are imported from other software.

Vectric's VCarve CAD/CAM software is included with most consumer and industrial CNC machines.

Vectric is the company that produces VCarve Desktop, VCarve Pro and Aspire. VCarve Desktop is a scaled-down version of VCarve Pro. It has much of the same functionality but is limited to a maximum job size of 24" (609.6mm) by 24" (609.6mm). It also does not share VCarve Pro's ability to automatically nest parts. At $350 USD, VCarve Desktop is worth your consideration if your machine's work area is within these 24" x 24" limits.

VCarve Pro has no limit on the size of the work or tool paths it can accommodate, and it includes powerful nesting capabilities to maximize material usage, so it often is provided with full-sized industrial CNC machines. Parts larger than your benchtop machine bed can be "tiled" or divided into smaller pieces that will fit in the work area. VCarve Pro also includes tool path templates that are not part of the desktop version. At $700 USD it is as capable as your desktop CNC will ever need.

Aspire is the next step up in CAD/CAM software from Vectric. It can do all the tool path tasks that VCarve Pro can, but Aspire is fully capable of creating CAD files in both 2D and 3D. This gives the speed of 2D CAD drawing for most of the work your CNC will do, while still being able to create 3D model forms when needed. Aspire is currently priced at $2,000 USD, so it's not for limited budgets. But if you need the added functionality, it's a good software option.

Vectric's products are purchased outright as opposed to the subscription model used by BobCAD and Fusion 360. But all of these CAD/CAM packages have free versions that you can download and try out before buying, and all three have excellent online training resources and tutorials.

PC or Handheld Control?

While you will be using software to write programs to run your new CNC, there will also be a control system for the physical operation of the machine. It calls up programs, sets a starting point, starts or stops a program run and similar functions. Some machines like mine connect directly to my computer with a USB cable. All the programs are stored on my PC, and there is a control program that actually operates the machine using my laptop screen and mouse. Other systems, like Axiom, operate without needing to be connected to your computer. Programs you write are loaded into the CNC's internal computer using thumb drives or SD cards, and the machine operations are controlled using a control panel or handheld pendant.

PC-based systems are typically a bit less expensive, since most of the storage and computing power come from a separate computer you provide. I have found that new programs often need some tweaking and adjustment when run for the first time. Having my PC right at the CNC makes these adjustments quick and easy, since I do not need to go to another computer in another room to rewrite the G-code. On the other hand, machines that run independently of being tethered to your computer could enable you to be running one program on the CNC while you are writing the next on your PC. The main drawback I find with hand-held pendants is that their screens display a limited number of characters, typically eight to 12. This means you need to keep program names very short with identifying characters at the beginning.

While you might be wondering about how woodshop dust could damage your computer if it's used with a CNC, I have never had a computer fail due to dust in more than two decades of CNC operation. Computers will get filled with dust and debris no matter where they are located, and I have found pet hair to be far worse than wood dust for clogging computers. Desktop computers can be easily opened to

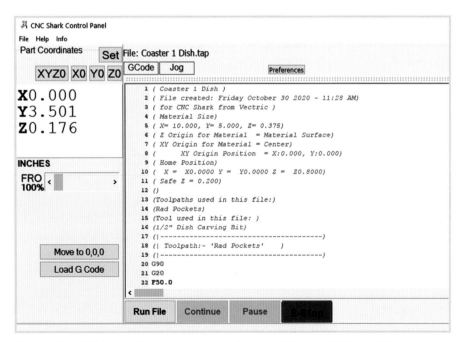

Here's the PC-based control screen I use with my CNC machine.

remove dust, and even laptops can benefit from dust removal. The key to cleaning your computer is to use a vacuum wherever possible, and do not use your shop air compressor to blow out dust. Use canned air meant for computers. It is a far gentler air stream than an air compressor and introduces no potential moisture to the electronics.

Spindle or Router?

When you are shopping for a CNC machine, you will notice that some use a handheld router motor to spin the router bit, while others use an integrated motor instead. These second types are known as "spindle motors" or "spindles". They are different from handheld router motors in a couple of ways: they are generally more powerful for the same size, they tend to be quieter and they can be more easily controlled by the computer. Let's look at the details.

Handheld router motors are rated as "periodic duty", meaning they are meant to run for a time, then shut off for a time. This is how handheld routers are generally used. On the other hand, spindles are designed to be "continuous duty" and can run for hours on end. CNC programs often will run for longer periods and even hours without stopping, so spindle motors are the better option. But they are more expensive. If you choose a CNC machine that uses a router motor, you'll probably need to replace the motor every couple of years depending on use, while a spindle will likely last a decade or longer.

My CNC Shark uses the Bosch Colt 1hp router. In the 12 years I have owned my CNC, I have replaced the Colt router twice. At about $99 each, this has not been an unreasonable return. Keeping the router clean and blowing out dust after each program run will help increase its lifespan. Additionally, if needed I can remove the Colt router from my CNC and use it as a handheld router. This isn't possible with a spindle motor. All else being equal, I would prefer to have a spindle motor, but the CNC Shark still meets my needs. So I'm happy to occasionally replace the router motor.

An Axiom CNC machine with a handheld pendant controller.

Online Communities

The issues above should help you narrow down your choices for your first CNC machine, but the decision can still be intimidating. The cost of a wrong choice will not be small. But as far as I know every CNC manufacturer of consumer systems has an active online user group or forum. These are excellent resources for you as you decide on a machine and even more so as you begin to learn how to use it. You can ask questions and get honest replies from owners of the machines you are interested in. These online communities also are able to provide helpful "armchair" tech support.

Whatever issue you may come across, it is highly likely that at least some members of the group have already learned the answers. The user groups I have been involved with are chock-full of members who enjoy helping each other out and sharing their projects. These online communities also exist for the CAD/CAM software. While you probably won't need to participate in all of these communities, once you've narrowed down your choices to two or three specific machines or software packages, the online groups can help you decide with more confidence.

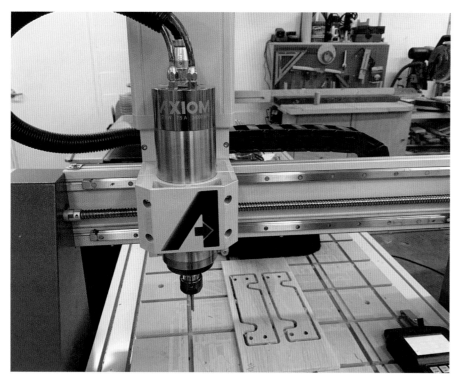

This CNC has an integrated spindle to drive bits rather than a router that must be installed in the machine.

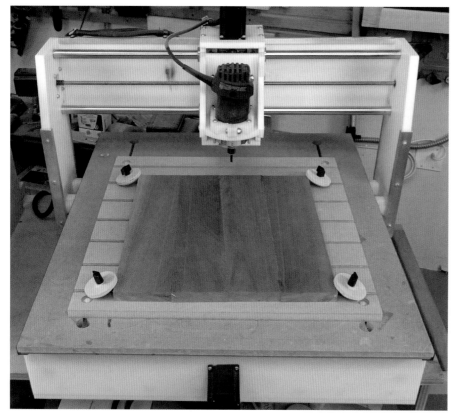

My CNC uses a Bosch Colt handheld router motor instead of spindle.

Alternative CNC Machine Formats

The vast majority of home shop CNC machines have a moving bridge that travels back and forth over a fixed bed, but this is not the only option. There are currently three unique machines that break this mold: the CarveWright, the Shaper Origin and the Maslow.

CarveWright

CarveWright was actually the first fully assembled home shop CNC to hit the market. I wrote a review of it for *Woodcraft Magazine* in 2007. At the time it was revolutionary, and it still is unique in many ways. To begin with, it is smaller than most typical CNC machines, about the same size as a benchtop planer. What stands out most is that the CarveWright has no moving bridge. The head travels side to side in the Y axis, but the stock itself moves front to back through the machine in the X axis. This configuration does limit the stock width to 14½" (368.3mm), but the machine is capable of milling stock up to 12 feet long. This is a length that few industrial CNC machines can accommodate, let alone benchtop versions.

As its name implies, the CarveWright is first and foremost a carving machine, designed from the beginning to create relief carvings. Its capabilities have expanded over the years to the point where there is now a rotary option allowing for carving all the way around objects up to 4¼" (107.95mm) in diameter. CarveWright's CAD/CAM software is also unique: it's available in specific modules so the user can purchase only the software needed, then add on capabilities as their skills grow. There are modules to work with 2D DXF drawings and 3D STL models. If carving and sign making are your primary CNC goals, the CarveWright system is worth a look.

The Carvewright CX Machine has the unique ability to mill stock up to 12 feet (3.6 meters) long.

An example of the type of work that can be done with the CarveWright CNC System.

Shaper Origin

If you have ever tried to make freehand cuts with your handheld router, you will appreciate the thinking behind the Shaper Origin CNC system from Shaper Tools. It's sort of a hybrid between hand routers and CNC machines. Shaper Origin consists of an outer frame a little bigger than a standard router base, which surrounds an inner frame that holds the router motor. You, the operator, move the machine by hand, following a visual of the tool path that's projected on a small video screen built into the machine. The CNC controls the inner frame, adjusting the movement of the bit to exactly follow the cut line and correcting the variations of your movement by up to ¼" in any direction. So you move the machine over the workpiece and the CNC motors augment your direction. The video screen on top also functions as a touch screen for setting up the machine. Its interface is designed to be easy to learn and use.

Shaper Origin knows its exact position on a workpiece and guides the cut through optics rather than the guide rails used by most CNC systems. The machine's cameras locate it by finding marks printed on special adhesive-backed tape that you lay out on the stock to be cut. This means that there is no real limit to the size of the stock that can be cut. One major advantage I can see for Shaper Origin is that it can be placed and used on parts that simply cannot be loaded onto any typical CNC machine. For example, mortises and other milling can be performed on countertops without needing to remove them.

Shaper Tools has recently released a new work station with the tape markings built in for the system to follow without needing to lay out new tape for each job. This will save users a lot of time and effort when working with smaller parts. For many hobbyists, having a CNC machine that can handle very large workpieces and still be stored in a cabinet when not in use is a very attractive package.

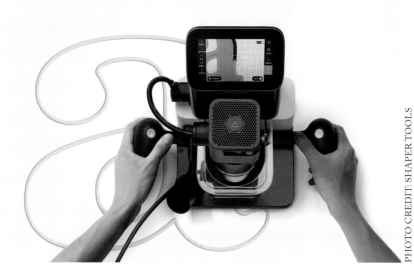

The Shaper Origin is a unique hand-guided CNC that automatically corrects the cut path as the operator moves the machine across the workpiece.

Shaper Origin's portable design enables it to mill large workpieces and to be brought to installations, such as countertops, that can't be moved.

Maslow

The Maslow CNC may be attractive to some readers for two main reasons. First, it's a system that is largely built by the user, with kits starting at under $400 USD. Second, Maslow is designed to cut full 4 x 8-foot sheet goods. The machine "bed" is oriented nearly vertical, so it does not require a lot of floor space for the size of stock it can accommodate. Maslow has been developed in an "open source" environment, which means its software and files are free for all to use and share.

The router on the Maslow is actually hung from a pair of cables to guide it around the panel being cut. It really does not look anything like any other CNC system. But it is able to cut large parts from full sheets of material, which no other system under $12,000 USD can do as far as I know. If your CNC dreams involve large parts, the Maslow system may be a solution for you.

These three alternative CNC formats really demonstrate that there is no one way, or even only a few ways, to accomplish a task. Machines with a fixed bed and moving bridge are the standard design for CNCs, but this configuration is not the only way to mill parts. These alternative machines each offer unique capabilities that may make them attractive, but be sure to look at the whole picture. The lack of a fixed bed makes locating and holding small parts like our luggage tag project more challenging, and two-sided milling as shown in the coaster project is likely to be impractical as well.

PHOTO CREDIT: MASLOW CNC

The cable guide system on the Maslow CNC allows it to machine full-sized sheet goods in a minimum of floor space.

PHOTO CREDIT: MASLOW CNC

Offering a spacious work area, the Maslow CNC system is an affordable solution that brings large-project capability to home shops.

Basic Operations

Home Positions

Not all CNC machines work in exactly the same way, but they are all pretty similar. The machines follow the instructions in the G-code, always moving from where it is at each moment to the position stated in the next line of code. All of this is done using coordinates where the start point on all three axes (X, Y, and Z) is set to zero. This is generally known as a **home position.**

So, where is home? That depends. My machine is older, so its home position starts out wherever the head is when the machine is turned on. Then I am free to move the head around and reset home to any place I choose. For most newer machines, however, all movements are based off a physical location. I call this the **hard home position**, because it cannot be moved or changed; it is mechanical. Every time the machine is powered up, the first thing it must do is find this hard home. The machine controller cannot assume that nothing was moved while it was turned off, so it will slowly move toward zero in all three axes until sensors register the position of the head.

Confused? Think of it this way: imagine you wake up in the middle of the night in your own room. It is pitch black, so you cannot see. You know where all the furniture is located through memory, but you are not fully sure where you are in the room. So, you stretch out your hands and move slowly until you gently touch a wall. Now you know where you are and can move around safely. The homing process is your CNC doing the same thing.

When your machine has a hard home, it can usually also store what I call **saved home positions**. These are positions you can choose and save within your control system. Usually, these will remain saved even if you shut the system down. They are very helpful for working with jigs and holding fixtures (one should be set to the corner of your spoil board once we have it set up later in this book). When you are using multiple programs on a single job, having them all using the same home position keeps them accurate, and with these saved homes, you can start any program from a known position.

In addition to hard and saved homes, there is also what I refer to as a soft or **temporary home position**. Your controller always shows the X, Y, and Z coordinates from its home position wherever the head is located. The origin may be the hard home or the saved home you are

working from. You always have the option of manually moving the head to a new location and resetting some or all of the axes to zero. All movements and programs will see this new position as a 0,0,0 and work from there until you choose a different home or shut down the system. This is called "soft" because it is not retained on shut down. Soft/temporary homes are highly useful when selecting the best section of a figured board to cut from, or for cutting parts out of odd-shaped scraps as you will see when we make our own clamps.

Defining the Process: CAD, CAM, G-codes, and Post Processors

Milling anything on your CNC requires three distinct steps. In step one, we begin by drawing the part and all of its details. This is the **CAD (Computer Aided Design)** portion. These are most often two-dimensional line drawings that contain no information about the depth of holes, pockets, or other features. For step two, cutting tools are assigned to the various lines and details in the drawings, with cutting speeds added and depths called out to tell the machine how to make a three-dimensional part from this two-dimensional drawing. This is the **CAM (Computer Aided Manufacture)** portion of the process.

The CAD and CAM can be separate programs, but most CNC machines come with programming software that can do both, known as **CAD/CAM software**. In the final step, all of the information worked out in the CAD and CAM processes needs to be translated into a format that the machine can read and follow. This is known as a **G-code** (see image on page 37), and the actual translation is done using a separate program called a **post processor**. Post processors are specifically tailored to machine brands and types. They allow for creating a working G-code for any number of different CNC machines from the same CAD/CAM information. Modern CAD/CAM software packages like VCarve Pro come with hundreds of post processors so they can be used with pretty much any **three-axis CNC**. In fact, I use VCarve to write programs for a number of clients with a wide variety of CNC machines. Once the basic design is done, I just need to use the correct post processor for the specific machine. This may all be a bit confusing, so let's walk through drawing and programming a simple

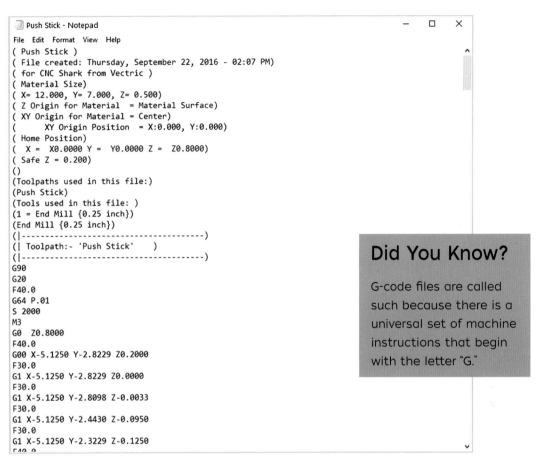

```
Push Stick - Notepad                                    —  □  ×

File  Edit  Format  View  Help
( Push Stick )
( File created: Thursday, September 22, 2016 - 02:07 PM)
( for CNC Shark from Vectric )
( Material Size)
( X= 12.000, Y= 7.000, Z= 0.500)
( Z Origin for Material  = Material Surface)
( XY Origin for Material = Center)
(      XY Origin Position = X:0.000, Y:0.000)
( Home Position)
(  X =  X0.0000 Y =  Y0.0000 Z =  Z0.8000)
( Safe Z = 0.200)
()
(Toolpaths used in this file:)
(Push Stick)
(Tools used in this file: )
(1 = End Mill {0.25 inch})
(End Mill {0.25 inch})
(|--------------------------------------)
(| Toolpath:- 'Push Stick'    )
(|--------------------------------------)
G90
G20
F40.0
G64 P.01
S 2000
M3
G0  Z0.8000
F40.0
G00 X-5.1250 Y-2.8229 Z0.2000
F30.0
G1 X-5.1250 Y-2.8229 Z0.0000
F30.0
G1 X-5.1250 Y-2.8098 Z-0.0033
F30.0
G1 X-5.1250 Y-2.4430 Z-0.0950
F30.0
G1 X-5.1250 Y-2.3229 Z-0.1250
F40.0
```

The G-code that instructs the CNC machine on how to mill parts can be read and edited as a text file.

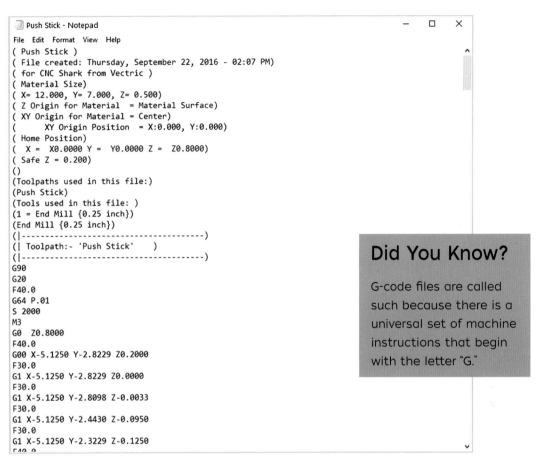

Did You Know?

G-code files are called such because there is a universal set of machine instructions that begin with the letter "G."

push stick, which is a safety tool woodworkers use to prevent injury by keeping their wood piece flat against the machine table or fence.

First you need a drawing of a push stick created in a CAD program. The drawing might be an existing **.dxf file** that someone else provides, or something you've drawn on your own in a CAD program (like DeltaCad®), or drawn inside your CAD/CAM program (see images, next page). At this point, you are working in only two dimensions, and all of the relevant points are located in both X and Y coordinates within the computer.

CAD/CAM programs typically have two user interfaces, one for drawing (CAD) and one for the toolpath work (CAM). VCarve Pro allows for both to be seen either together or one at a time. I like to view one at a time to maximize the size of the work area on the screen.

Once the two-dimensional drawing is completed, the CAM interface is accessed so instructions can be input, depending on the types of machining to be done, tools to be used, depths of cut, and other important information. These are generically called **toolpaths**, and they are divided into types that help decide methods of work. Cutting steps are managed differently than steps for milling pockets, drilling holes, or engraving and carving.

By applying depths of cut to the toolpaths, the two-dimensional drawing becomes three-dimensional, as it assigns depth to the various cutting instructions. The drawing guides the bit in the X and Y directions to trace the shape of the push stick, and the depth defined in the toolpath controls the Z axis. These three axes, X, Y, and Z, are known as **Cartesian coordinates**. Cartesian coordinates are used in milling machines like the CNC as well as in mathematics for graphing. Depending on your machine, the X axis may be front to back and the Y axis side to side, or X side to side and Y front to back. The Z axis is always up and down, controlling the depth of cut.

Because the push stick is a simple cutout with no internal details, you only need to fill in the information for one toolpath. In VCarve, cutting to lines is called a **profile toolpath**, but other programs may use slightly different terms. Modern CAD/CAM programs typically offer a very good visual simulation that clearly shows the results of the toolpath choices. The image on page 39 (top) shows the push stick profile cutout, but with three small red "tabs" left behind to hold the part as it is cut out.

Once you are satisfied with the results of your programming, the information must be translated into a format that the CNC machine itself can read—the G-code previously discussed. You can choose the correct post processor contained within your CAD/CAM program and save the resulting file where you choose. If your machine is a common brand, the right post processor is probably included in the CAD/CAM package you own. If not, the machine manufacturer can provide you with the required post processor file for

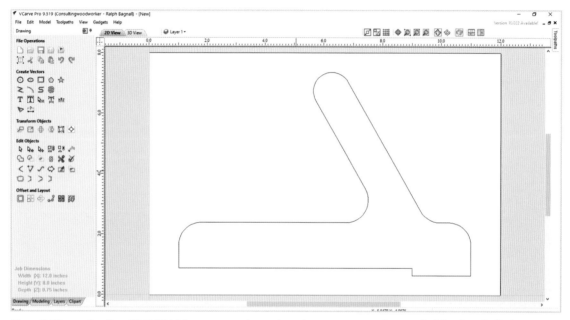

The CAD part of a CAD/CAM program has drawing commands available for defining the project.

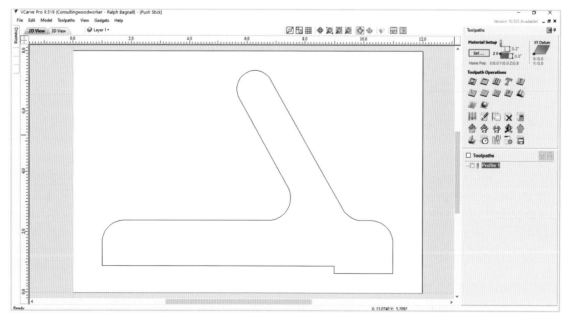

The CAM side of the program above showing the toolpath commands available for milling the project.

your machine, which you can then load into your CAD/CAM program.

CNC machines are really quite primitive as far as computers go. The controller that actually runs the machine cannot interpret all of the information contained in the CAD/CAM program you just completed. The G-code that will actually guide your CNC's movements is just a simple text file. The post processor is a translation program that takes all of the data within the CAD/CAM program and rewrites it into a G-code file that is tailored to your machine (see image on page 37). Even for a simple cutout like the push stick, this text file contains literally hundreds of lines, each a specific movement of the head from place to place. This is the part that the machinist needed to type in by hand in the past, before CAD/CAM software and post processors were invented.

If this all sounds too confusing at the moment, don't worry. We are going to cover all of this and more in much greater detail as we go. Just keep this overview of the process in mind as you read through the rest of the book.

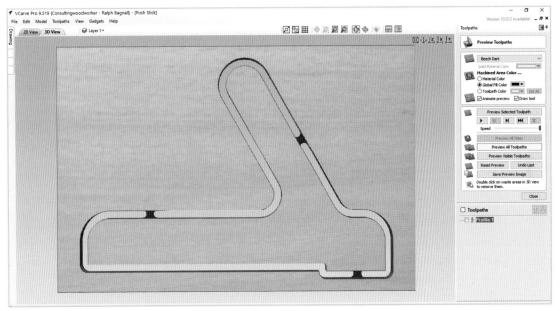

A graphic preview of the push stick project as rendered by the CAD/CAM software.

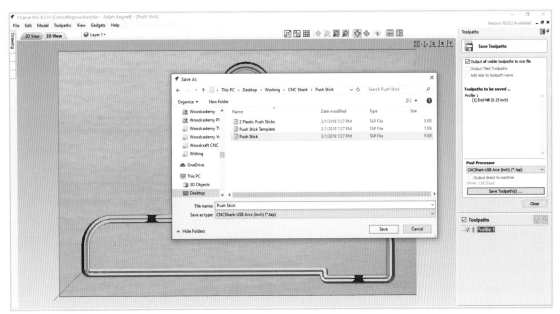

This image shows a G-code file being saved. The post processor has been selected for my CNC Shark machine.

Choosing Bits for Your CNC

PHOTO CREDIT: INFINITY TOOLS

A wide range of router bits are now made specifically for use in benchtop CNCs, but most of your "standard" router bits can be used as well.

One of the first questions I get when teaching new CNC users is about which bits can be used with their machine. The short answer is that you can pretty much use the same bits as your handheld router and router table.

While there are specialty bits in the industrial market made specifically for CNC use, these are generally designed to perform at feeds and speeds far faster than any of these benchtop machines are capable of, so they are unsuitable for our benchtop machines. Spiral bits are an exception; having initially been designed for CNC use, they have become popular for handheld routers as well. Excepting bearing guided router bits, the line between "CNC Bits" and "Router Bits" is nearly irrelevant now.

Your CNC cannot use bearing guided bits like your handheld router can, but pretty much any router bit without a bearing can be used on your CNC machine. In your handheld router, the bearing rides along the edge of the part, with your arms acting as "springs" to control the contact and follow the edge. This is how the cut is controlled. On your CNC, you do not need to use the bearing for control, the machine frame is controlling where the bit goes, and a bearing will simply be in the way and cause problems. There remain many styles and types of bits you can use with your CNC machine, so let's look at some specific bit types and consider reasons you might choose one over another.

A selection of router bits that I have used with both my handheld router and CNC machine.

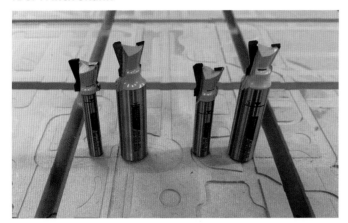

Most router bits are available in both ½" (12.7mm) and ¼" (6.35mm) shank sizes. This dovetail bit from MicroJig is also available with 8mm and 12mm shanks to fit European routers.

My CNC Shark uses a Bosch Colt router for the motor, so I have no choice but to use ¼" (6.35mm) shank bits. If your system can use ½" (12.7mm) shank bits, you should whenever possible. A ½" shank bit is not twice as stiff as a ¼" shank, it is actually eight times stiffer, which means much less vibration and cleaner cuts at higher feed rates. The only exception I have found to this rule of thumb is when using a ¼" diameter or smaller bit. To make a ¼" bit with a ½" shank means the body must be turned down on a lathe, and the neck where this transition is made is always the weak part. I have broken a number of ¼" bits with ½" shanks, but very few ¼" bits with ¼" shanks. Those using metric bits will generally need to choose between 8mm and 12mm shank bits. The same rules apply, so use the bigger shank where possible.

Carbide Tip and Solid Carbide

Many bits are made as a steel body with carbide cutting edges braised on, and others are milled completely from a single piece of solid carbide. In many cases, the same bit may be bought in both configurations. There are technical differences, but honestly, at the feeds and speeds our benchtop CNCs are operating at, there is no specific benefit to either type. The ⅛" (4mm) and ¹⁄₁₆" (1.5mm) bits we'll be using later in the book are solid carbide because it is impractical to try and braise such thin carbide pieces to a steel shank. But many ¼" (6mm) and ½" (12mm) straight bits, V bits and fluting bits can be had in steel body or solid carbide. Steel body bits tend to be a bit less expensive. Buy whichever is available in the size/shape you need at the best price when you buy it.

Router bits usually consist of a steel shank and body with carbide tips brazed onto it, but bits also can be milled from solid carbide. These two straight bits are made differently but make the same cut in your CNC.

Spiral or Straight Bits

Spiral bits are good general-purpose bits, and they can cut at higher feed speeds than an equivalent straight bit. In many materials, they provide a smooth cut. The curve of the spiral bit's cutting edge has the same effect as angling a hand plane or chisel during a cut; the angle creates a shearing cut like a knife rather than a straight chopping cut like a chisel. This can be a great advantage allowing for higher feed speeds with lower RPM to avoid friction and heat.

While one of these bits is spiral ground and the other is not, both make the same flat-walled, flat-bottomed cut when mounted in your CNC. In plywood, the straight bit on the right typically makes a better cut.

Choosing Bits for Your CNC **41**

Spiral bits aren't always the best choice. They will cut plywood quickly, but the edge quality of the cut will be much less smooth than a simple steel-bodied straight bit with carbide cutters. The spiral bits tend to perform better cutting with the grain and less well cutting against it. So plywood edges cut with spiral bits tend to be fuzzy between layers where the grain direction changes. When cutting plywood, the slower-cutting straight bit will save you time sanding the edges of the parts after milling.

The many cutting edges of the end mill shown here work well with metals, but in wood the extra cutting edges generate excessive friction. Benchtop CNCs work better with one- or two-flute bits than end mills.

Metalworking CNCs use liquid to flood the work area and keep stock and tools from overheating. Obviously, this is not possible with a woodworking machine. The one or two cutting edges on a woodworking bit create far less friction than the many edges on end mills, and the higher feed speeds used in woodworking also keep the heat under control. We will discuss the issue of overheating throughout the book. In general, stick to woodworking bits in your CNC machine.

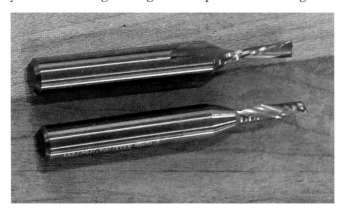

An upshear spiral cutting edge twists up from tip to shank, while the downshear twists from shank to tip. Even though both make the same size cut, the upshear removes chips better from deep grooves and pockets while the downshear provides chip free cutting with veneers and laminates.

Spirals can be had in both "up" and "down" configurations. The up or down is based on the direction of the spiral twist relative to the tip. An up spiral bit will pull chips up out of the kerf being cut much like a drill bit. A down spiral pushes the chips toward the stock being cut. The lifting cut of the up spiral is useful for helping to clear chips in deep slots, but it also tends to tear out the fragile face veneer layer on plywood. When I'm cutting shallow pockets and dadoes in plywood, I use a down spiral. The spiral cuts down into the panel from above, so the fragile veneer layer is supported by the substrate as it is cut, leaving the top edges of the cut exceptionally clean and smooth.

Machinist's End Mills

Many CAD/CAM programs refer to straight bits as "end mills". This is a holdover from the metal machining industry where CNCs got their start. While typical woodworking bits are technically end mills, they are not interchangeable with end mills made for cutting metals. The feeds and speeds needed for woodworking are simply not compatible with metal cutting tools.

Profile Bits

Most profile router bits are bearing guided and unsuitable for CNC use, but many can be found without bearings. These can greatly expand your CNC's capabilities. V bits, dovetail bits, round overs and ogee bits are just a few of the profiles available. Using these alongside your straight bits allows for carving text and images into a sign, adding a nice profile to the edge, and cutting it out all on the CNC. Creative use of profile bits with your machine will take your projects to another level.

Both of these bits cut the same profile, but only bits without pilot bearings can be used in your CNC machine.

Buy the Best You Can Afford

A surprising number of the router bits in a woodworking collection can be used in a benchtop CNC machine as well.

Here is my advice for buying any tools or bits: there are many discount bits on the market, but you are not saving money if the cut quality is poor or if you have to replace them more often. All of the reputable router bit brands you prefer for your handheld router should be your go-to source for CNC bits as well.

Many manufacturers are now offering sets of bits for CNC users. I have always advised caution when buying bit sets for handheld routers, and the same goes for these CNC kits. A set that includes a couple straight cutters in common sizes, V bits in 90 and 60 degrees and a round nose or two is a good place to start, but a set of 10 or more bits always seems to mean that a few get used regularly, and a bunch more that just gather dust. I generally prefer to buy bits as I need them. It seems to save money over the long term.

The bottom line is that using a clean, sharp bit at the proper feed rate and RPM will provide the best results on your CNC just as it will with your handheld or table router. I keep all my router bits in one dedicated cabinet and often use the same bits for handheld routing or CNC work.

Bit and Collet Maintenance

A little-known fact among woodworkers is that carbide bits and blades are rarely worn out through use. What we call "getting dull" is most often the carbide being damaged by heat. Carbide is not a solid metal like steel but rather is made up of carbide particles bonded together with another metal, often tungsten. Excessive heating of the carbide tips causes the binder to break down, releasing the carbide particles. This happens first at the cutting edge where the carbide is thinnest and heats fastest.

Your CNC machine gives you control over the feed and speed that bits are run at, so you can prevent this type of heat damage through proper programming of the tool paths. Ramping into your cuts, moving the bit as fast as possible and keeping the RPM as low as possible are all techniques I talk about throughout this book. But keeping your bits and collets clean is just as important for getting the best tool life and the highest-quality cuts.

Your bits and collets require regular cleaning and maintenance just as your CNC machine does to provide you with the best possible performance.

Choosing Bits for Your CNC

The bit on the right has been damaged through excessive heat during cutting. It is beyond repair and should be recycled.

Cleaning some bits. Cleaning stations like this one from MicroJig can help you spend less time cleaning and more time cutting.

When a bit is cutting, the heat that develops tends to vaporize the sap or resins in solid woods and the glues in sheet goods. This vapor will condense onto the first cooler surface it touches, which will usually be the back bevels of the carbide, the bit shank and the collet. This is the "gunk" that you see appearing on your bits and blades when you are using any tool. As these burned-on deposits build up behind the cutting edges, they start to rub against the stock being cut and add to the friction heat that is damaging your bits and reducing their cutting performance. So keeping bits clean is crucial.

Many commercial cleaners are formulated specifically for woodworking bits and blades, and fortunately many of them are made from organic and non-caustic bases so they are much safer to use in the home shop. Some use citric acid from oranges and lemons to clean, and I have had great success with soy-based products as well. Regardless of what cleaning product you choose, what's important is to clean your tooling regularly.

I often start with a razor blade to remove the bulk of the build-up, then soak the bits in the cleaner. They can be scrubbed with a brass wire brush, steel wool or even non-woven abrasives like Scotch-Brite™. Don't forget to inspect and clean your collets too. If there is build-up on the shank of your bits, then the collet also will be dirty. Dirty collets can lead to excess vibration of the bit and can even cause the bit to slip out of the collet, potentially damaging parts and machine.

Testing the sharpness of a router bit by feeling the edge with your finger or thumb is simply not a reliable gauge. "Sharpness" is not only about the cutting edge but also the clearance bevels of the blades, and those are much more difficult to test. If your bit is clean and cuts poorly or makes excessive noise when cutting (compared to a newer bit) then it has likely come to the end of its useful life span and needs to be replaced. Except for physical damage to the carbide or overheating, you can expect to get many hours of actual cutting time from your bits on the CNC.

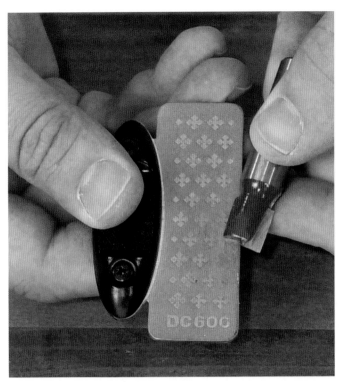

You can use a diamond plate to hone the flat face of the carbide cutters on your bits. This can refresh the edge and extend the bit life.

Collets are not designed to last forever and should be replaced every two to three years depending on how often you use your CNC.

If the bit has not been burned or the cutting edge chipped, honing the face of a bit can refresh the edge and extend the life of the bit. A diamond sharpening stone or card (see photo above) will work quite well on carbide. Just hold the bit with the carbide face flat to the stone and rub it back and forth, keeping the face firmly pressed against the stone. You do not want to try and sharpen the carbide's beveled edge or change the shape of its straight or curved profile in the process. You only want to hone the flat face of the carbide. This may seem wrong, but by flattening the face, you restore the cutting edge with little or no change to the bit diameter or profile shape.

One final note here: router collets do not last forever. The inner sleeve inside the nut that holds the bit (see photo above) is made of spring steel so that it can compress to clamp your bit in place. The heating and cooling that they go through with each use eventually reverses the heat treating they got at the factory to make them flexible. With daily use, router collets (CNC or handheld) should be replaced every year. In home shop use, consider replacing collets every two or three years depending on how often you use them. The recommendation from collet manufacturers is around 500 hours of use.

CHAPTER 2

BASIC FITTINGS

Most CNC machines use a bed to support the stock being cut and a bridge above that guides the router. The Carvewright and Shaper Origin CNC machines are exceptions (see Alternative CNC Machine Formats, page 33), but most machines use this format and come with a couple of clamps for securing the stock. Different CNC tasks require different setups and clamping techniques, and the machine bed needs to be protected when the bit is cutting through the stock. Some projects like the Custom Coaster Set in Chapter 5 will need dedicated holding fixtures, but over the years I have developed a bed board and clamp set that will quickly and easily secure most jobs while protecting the machine from cutting into itself. Best of all, making these accessories for your shop is a great way to get started using your new CNC.

Making a Spoil Board

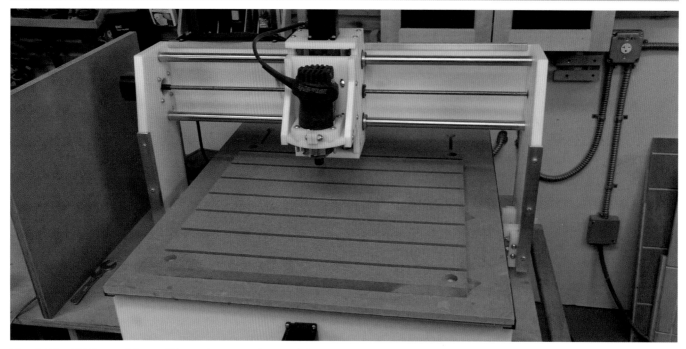

Fig. 1: My CNC machine with a shop-made spoil board attached.

The first project you will be making with your CNC is a **spoil board** (Fig. 1). It will be used in just about every future project you will do on your machine. The design of this spoil board provides three critical advantages to your CNC:

1. It allows for cutting all the way through the stock without damaging the factory bed of your machine. It is sacrificial, which is why it is called a spoil board.

2. You will be machining slots into it, so securely clamping stock to it is easy and quick.

3. The top of the spoil board will be milled flat using the machine itself, which ensures that the work surface is a highly accurate Z-axis reference point so the top of the spoil board is a known position. This is especially important for tasks like milling hinge pockets or engraving fine details.

I have always used spoil boards on all flatbed CNC machines, even large industrial machines that cost more than $100,000 (€84,500). These handy boards save time, money, and material.

Supplies

◇ ¾" (20mm) MDF at least 1" (25mm) smaller than the working envelope of your CNC

◇ ¼" (6mm) straight bit

◇ ²⁵⁄₆₄" (10mm) keyhole bit (similar to Freud #70-104)

◇ ¼" (10mm)-20 (M6 x 1mm) nylon bolts with washers

◇ ¼" (10mm)-20 (M6 x 1mm) T-nuts, to fit your CNC table

Other Wood Options

MDF (medium density fiberboard) is the preferred material for spoil boards, although I have had excellent results using some pre-glued softwood panels. MDF is inexpensive, easy to mill, and is more dimensionally consistent than plywood. Plywood is stronger, but since the spoil board will need to be milled smooth regularly, a plywood surface will not remain smooth as the layers (plys) are cut through. Using MDF will mean a smooth surface each time the spoil board is planed down. The board needs to be sized to fit within the actual **working envelope** of your machine so that the top surface can be milled flat and smooth. This requires the router bit to reach all four corners of the spoil board and a little bit beyond. On my machine, 22" x 22" (560 x 560mm) is a good size.

Determining the Working Envelope

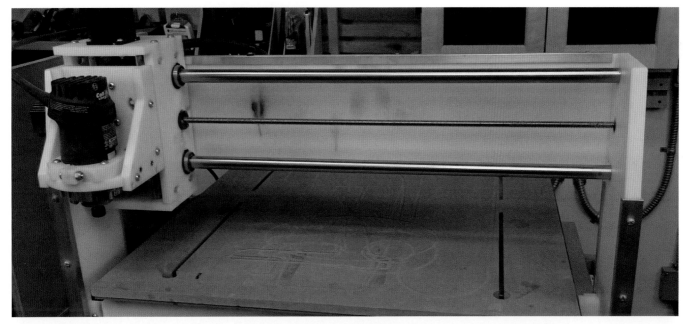

Fig. 2: Moving the machine head to set the working zero point.

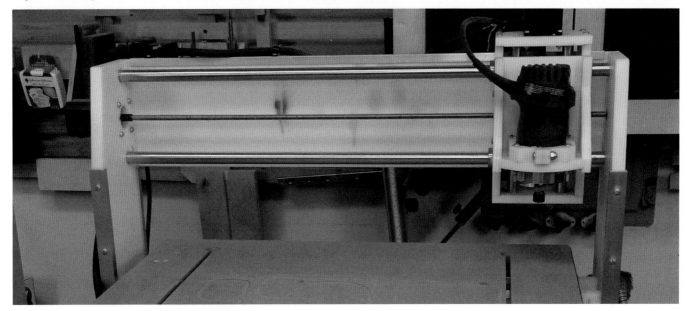

Fig. 3: Moving the machine head to find the maximum head travel.

1. The body and frame of a CNC machine obviously have to be larger than the stock being milled to allow room for the head to move front to back and side to side. The area within these limitations is the **working envelope** of your machine. This may not be exactly the same as the specified size of your machine.

2. If the working envelope is not specified in the machine instructions, you can simply move the head from corner to corner, recording the distances. Start by "manually" moving the head using the computer controller as far into the front left corner as possible. Check your machine instructions for the manual moving procedure (Fig. 2). Set this spot to zero in both the X and Y axes. Then move the head as far into the back right corner of the machine as possible (Fig. 3). The X and Y readouts are the extent of your working envelope. On my machine, it is a bit more than 22" (560mm) in both X and Y, so I use 22" x 22" (560 x 560mm) as the working envelope of my machine.

Surfacing the Spoil Board

3. The spoil board first needs to be cut to size and mounted onto your machine's factory bed. I prefer using nylon bolts for mounting (Fig. 4). If I make a mistake and cut across the bolt, it won't damage my router bit. Just measure and mark hole locations in the four corners of the spoil board that line up with whatever mounting system forms the bed of your machine, counterbore, and drill through holes. You can program these in, but it is just as easy to simply drill the mounting holes by hand or on the drill press.

Fig. 4: Nylon bolts are ideal for mounting the spoil board. They will not damage your router bit if you accidentally cut into them.

4. Once the board is mounted within the working envelope, the first step is to machine it flat. Notice that I did not say *level*. In this case, flat is really parallel to the plane that the bridge (the machine section with the router motor and bit) travels. We want to have a working surface that has the same Z-reference height across its entire surface. In other words, if you set the bit to 0.001" (0.03mm) above the work surface at any point, the head can be moved to any other point on the surface and still be 0.001" (0.03mm) above the work surface. We will see later, especially when engraving letters in signs, how important it is to have a known working surface.

5. When you machine it flat, this process is called a **table mill** or **surfacing**. It's programmed as a very shallow pocket toolpath that covers the entire spoil board. Since this is likely the first program you will have written, let's walk through the process of creating the pocket tool path you need. Begin by

Fig. 5: The table needs to be milled with a flat-bottom bit. The larger dedicated bit with carbide inserts (left) will mill more quickly than the standard straight bit (right) but both will get the job done.

drawing a rectangle the size of your spoil board, plus about ½" (13mm) on each side. This is done on the CAD side of your software. The rectangle is all you need for this step, so switch over to the CAM side of your program.

The reason for adding the extra ½" (13mm) to the spoil board size is because a round router bit cannot cut a square inside corner — it will always leave a radius. If you tell the machine to mill within the perimeter of the actual spoil board, the bit will not go past any edge, and it will leave material in each corner where the cutter doesn't reach. The extra dimension in your pocket rectangle ensures the corners get cut too.

Next, select the lines (vectors) that you want to tool path by clicking on them. They will highlight to show the selection. Now choose the tool path needed: in this case, that's the "Pocket Toolpath". A new window will open with data blanks that need to be filled in. This is where the 2D drawing gets depth added to it. The first data to enter is the depth of the cut. We just want to shave the top of the spoil board flat, so we will enter ¹⁄₃₂". Interestingly, since the fraction ¹⁄₃₂ is actually an equation meaning "1 divided by 32", you can enter the depth of cut as "¹⁄₃₂" and it will change to "0.0325" as it is entered.

6. The table mill is machined with a flat-bottom bit. I use a ¾" (20mm) bit because my CNC's router can only take ¼" (6mm) shank bits. With a ½" (12mm) collet router, you can mill with larger-diameter bits. The larger the tool diameter, the faster the surface can be planed, and the sooner you can get to work. There are even special tools designed to mill the table quickly and accurately (Fig. 5).

The program next asks you to choose how the pocket will be cut. You can choose it to cut from the center outward or working side to side. You can also choose whether the bit will make a climb cut or a conventional cut. In this case, MDF has no grain, so which cutting direction you choose does not really matter.

The next selection to make is whether to use "Ramp Plunge Moves" or not. Larger flat-bottom bits are often not designed for plunging cuts. Ramping the bit in the program will move the bit sideways as it lowers to the cut depth, so the bit will move downward without issue.

Finally, choose a name for the tool path and click on the "Calculate" button. You will see the tool path appear on your stock, and you can run a simulation to see how it will actually cut on your machine.

Select the tool path and save the G-code. This is done in various ways depending on your software. In VCarve Pro, returning to the main CAM screen after calculating the tool path allows me to export the G-code.

7. You can now start the program run. The G-code program is either selected or loaded into the control software depending on your machine. The exact process for starting the program run will be found in your machine manual. It is usually a "GO" or "OK" button that can be selected once you are ready. As you get ready to press the "GO" button, think about what the program will have the machine do first. Where will the head move with the bit to make the first cut? Knowing where you think the machine should be headed will allow you more quickly react if things should go in the wrong direction. Every machine has at least "Pause" and "Stop" buttons in the software, and most have a physical "Emergency Stop" button, any of which can stop the program run safely if there are errors.

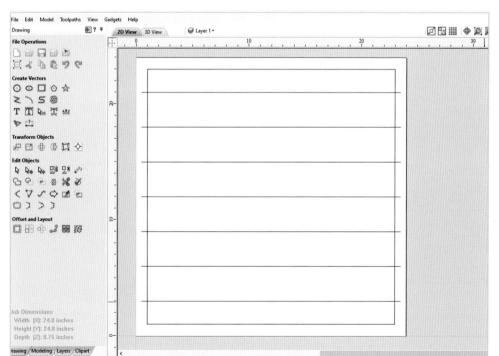

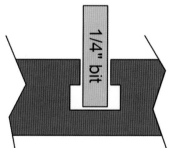

Fig. 7: The T-slot is positioned to completely over-cut the smaller relief.

Fig. 6: This spoil board drawing shows the job size, the perimeter of the spoil board, and the T-slot locations.

Milling the T-Slots

8. A series of T-slots need to be milled into the spoil board in order to efficiently clamp parts in place using T-bolts and other appropriate hardware. This is best done on the machine because you just need to mill a series of slots. Open your CAD/CAM software and draw a rectangle the same size as the spoil board. Draw in lines that represent where the T-slots will be located (Fig. 6). On my machine, I have the initial line set 2" (50mm) from the front edge and a series of parallel lines spaced 3" (75mm) apart, which on a 22" (550mm) surface provides even spacing. The 3" (75mm) measurement is not critical, but I have found that this dimension works well, especially with the clamping disks we will be making later to use with the T-slots (see pages 57-61).

9. Each T-slot is actually formed as an upside-down "T." The sort of router bit that can cut this is referred to as an **undercut bit**. I always recommend milling a relief groove before using any undercut bit. The opening in the groove is narrower than the widest part of the slot, so chips and debris get trapped in the groove. Removing most of the waste with a relief groove means less debris in the T-slot to begin with and a path for the material to flow out. Using relief grooves will greatly improve the bit life by helping reduce friction.

10. You need to program the relief grooves along the same line that the T-slot will be cut. The relief groove and the T-slot will share the same path, but the T-slot will be formed by two passes side by side where the relief groove only uses one pass. A single line in the drawing can be used to tool path each relief groove. This groove will be milled away as the T-slot is formed, as shown in Fig. 7. It is only there to remove some material before the undercut is made, as described above. In this case, you want to program a ¼" (6mm) groove to be cut 5⁄16" (8mm) deep. Select the groove lines on your drawing, click on the "2D Profile Toolpath" button and fill in the data in the window that appears.

To set the cutting depth to 5⁄16" (8mm), enter .312" (8mm) where it says "Cutting Depth". Now select the ¼" (6mm) router bit (also called an end mill), and set the cut to be "On" the vectors, which means the center of the bit will exactly follow the line on the CAD drawing.

There's no need to ramp these cuts, since they will start and end beyond the edges of the spoil board. The "Add Tabs to Toolpath" can be left alone as well, since there will be no through cuts. The final prompt will ask you for a file name. The software will default to something like "Profile 1" or "Pocket 2" as needed, but now is the time to begin naming all your tool paths. Do this with all programs, even the simple

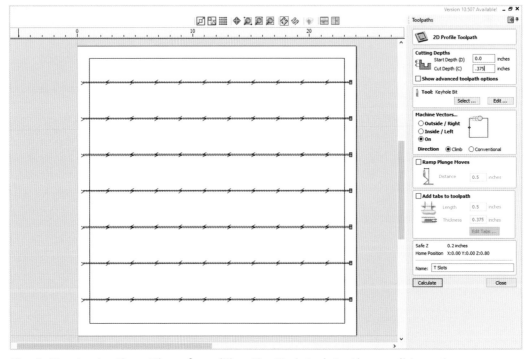

Fig. 8: The toolpath settings for milling the T-slots into the spoil board.

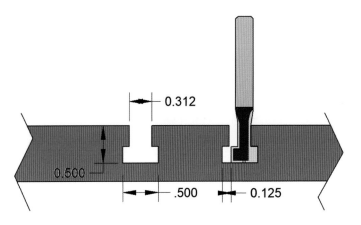

Fig. 9: This graphic shows how the T-slot is milled to final width using a smaller-diameter bit.

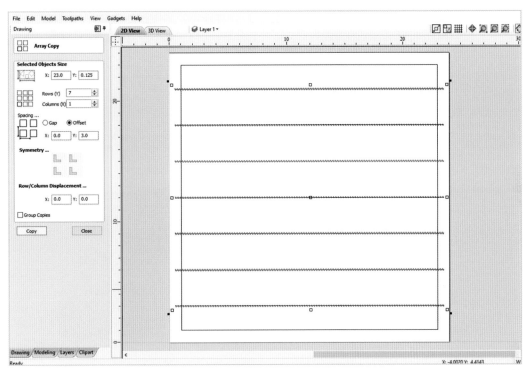

Fig. 10: Multiple T-slot paths can be added to the drawing using the array function.

ones, to build good habits. On complex programs that use multiple tool paths, having names like "Profile 1", "Profile 2" and "Profile 3" make editing far more difficult and can lead to mistakes. It's easier to build good programming habits than to break bad ones.

Finally, click on the "Calculate" button and the tool path will be created for you with all X, Y and Z coordinates figured out automatically in the software. You can now use the preview feature (see page 74) to see what the cuts should actually look like when you cut your spoil board. There ought to be a series of ¼" (6mm) wide lines running side to side.

11. The next step is to draw in and program the T-slot cuts. These will use different lines within the drawing than the relief grooves, but one of the cool features of CAD/CAM software is that you can pick and choose which parts of the drawing you are working with at any given time. So rather than have separate drawings for the relief grooves and the T-slots, everything can be worked on from a single CAD drawing. This saves time and reduces the opportunity for errors. We can see in Fig. 7 that the T-slots and relief grooves share a centerline, but the T-slots will overcut the relief grooves, removing them completely.

Making a Spoil Board   **53**

Fig. 11: The relief groove being milled into the spoil board using a ¼" (6mm) straight bit.

Fig. 12: This photo shows the first half of the T-slot being milled along the path of the relief groove using the keyhole bit.

12. The T-slots will be cut using a keyhole bit (Fig. 8). It cuts a T-shaped slot, but not as large as we need. There are no ¼" (6mm) shank bits designed to cut the exact slot we want, but the CNC makes it very easy to create the exact size needed by making two passes. Programming this requires no resetting of fences that would be needed to do this with a hand router. The Freud bit we are using is sized so that two passes, ⅛" (3mm) apart, will cut a T-slot that can be used with both ¼"-20 (M6 x 1mm) and ⁵⁄₁₆"-18 (M8 x 1.25mm) sizes of T-bolts (Fig. 9).

13. Since the relief grooves and T-slots share a centerline, the easy way to draw the two T-slot lines is to work off of the relief groove lines. Your CAD/CAM software allows for copying and pasting. So copy the first relief groove line, and paste a copy of it ¹⁄₁₆" (1.56mm) to one side of the original line. In VCarve, the copy appears on top of the original line and is already selected. It can be moved using "Move Selected Objects" in the CAD area. With this line still selected (it should still be highlighted) offset a second line ⅛" (3.2mm) from the copy to the other side of the original relief groove line. These two new lines should straddle the relief groove line.

Draw a short line segment between the two new lines at either the left or right ends, it does not matter

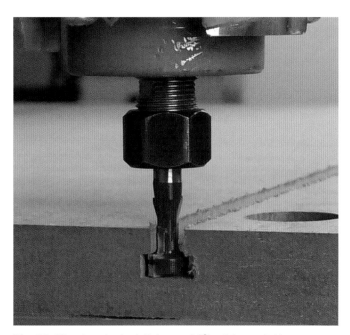

Fig. 13: The keyhole bit has shifted to cut the second half of the T-slot, completely overcutting the relief groove.

which. This will give you a long "U" shape on its side running side to side across your spoil board. This is the path that the T-bit will travel across the spoil board, then it will shift over by ⅛" (3.2mm) and run back across the spoil board to create the proper width T-slot, as illustrated in Fig. 9.

14. You could do the same thing for all of the grooves on your spoil board, but pretty much all CAD/CAM programs allow for copying items in an **array**. This command is one you want to become familiar with. Arrays allow for precisely laying out copies of drawing items in lines, columns, grids, and even circular patterns. It can save a lot of time when drawing (Fig. 10).

15. Creating the toolpath for the slots is simple, but you must first add the tool to the machine database. Add it as an end mill or straight bit with the actual diameter of the bit entered. Mine measures $^{25}/_{64}$" (10mm). Set the pass depth at ½" (13mm), because as an undercut bit, it should only cut at the depth set in the design; it must not make multiple passes to reach final depth as the ¼" (6mm) bit did when cutting the relief groove. When choosing the vector offset, use "On the Line." Most vectors we use will be **closed vectors**, where all parts of a detail are connected. This makes it easy for the program to choose the outside or inside of the vector and offset the bit to cut to the line, but with the open "U" shape, inside and outside may not be clear. Cutting on the line eliminates any potential issues. This is why you created the lines ⅛" (3mm) apart.

16. To begin milling the spoil board, mount a ¼" (6mm) straight bit into the collet and set the Z-axis according to the instructions for your CNC machine. The Z-axis should be set to the top of the newly flattened spoil board. The X and Y start point should be the same as used for the table mill program. But if the machine has been turned off, the spoil board moved at all or the home reset, you may need to reset the start point to the left corner closest to the front of the machine. Move the head using the controller until the bit is exactly centered over the corner of the spoil board, and reset X and Y to zero according to your machine's manual.

Load the relief groove G-code program you created earlier into the machine, and if everything is ready, run the program. As mentioned before, be thinking about where the bit will go first so if something different happens, you can pause the run to avoid problems. The program will cut a series of slots across the spoil board, but it might not cut the front

one first and move in sequence to the back, so do not panic if the first groove cut is not the first in line. I recommend that you watch the preview in the CAD/CAM program, because it will show which grooves will be cut in which order so you'll know what to expect.

When the program finishes running, most machines will raise the bit to a safe height above the spoil board and simply stop where it is or return to the starting position in the X- and Y-axes, keeping the bit above the stock. At this point, change the bit to the T-slot cutter and load the next program into the controller to run it next.

17. Mount the keyhole bit into the router collet and reset the Z-axis height according to your machine instructions. There is no need to reset the X- or Y-axes, in fact they must remain the same. Run the table mill, relief groove, and T-slot programs from the same start position, only resetting the Z-axis with each change of bits. This will keep everything aligned. Most CNC machines allow for a number of start positions to be saved as home positions, but be aware that on some machines you may encounter issues when changing the start position while using multiple programs for the same project.

18. The first pass will form half of the T-slot (Fig. 12), and the second will finish the profile. If programmed and run properly, cutting of the T-slot should completely remove the relief groove (Fig. 13). While having a single bit to cut the T-slot in one pass may seem more convenient, using a smaller bit to create the exact size feature you want is much more flexible. If the keyhole bit used here should ever be discontinued, or you decide you want the T-slot a bit larger or a bit smaller, the program can be easily adjusted to make the changes without starting over from scratch. Running the T-slot program will only take a few minutes to mill, and when the spoil board needs replacement, all you will need to do is cut a new piece of MDF and run the three programs above to create a fresh one.

Alternative Slot Options

Fig. A: Here are three different bits that can be used to form T-slots: T-slot bit #143-0702 from Eagle America (left), Rockler T-slot bit #26099 (center), and Freud keyhole bit #70-104 (right).

Fig. B: A similar spoil board that uses dovetail grooves and hardware instead of T-slots.

I have been using T-slots on spoil boards for decades, and they have always served me very well. In the past, I used a keyhole bit to create the slot I wanted, but there are some alternatives. Rockler makes a very good T-slot bit (Fig. A) that cuts the right-width slot, but because there is no cutter on the neck, a ⅜" (10mm) slot must be milled first. This won't be an issue if you mill a relief groove using undercut bits. I have used this type of bit to make spoil boards for my CNC, but because it has a ½" (12mm) shank, I cannot use it in my machine. Whiteside Machine Company makes a T-slot cutter that needs a precut slot at least ¼" (6mm) wide, and it will cut a slot ¾" (19mm) wide. Depending on your T-slot hardware, this may or may not be too wide.

Some machines come with MDF strips attached to the aluminum T-track machine bed. This works just fine for holding and eliminates the need to make your own spoil board. There is, however, always a trade-off. This system requires longer T-bolts, and the MDF strips will need replacing at some point, which cannot easily be done right on the machine. On the other hand, you can begin through-cuts on your machine right away.

MICROJIG® offers a hardware set that works in a ½" (13mm)-wide, 14° dovetail groove. I have made a spoil board using this system, and it works just fine (Fig. B). Any brand of ½" (12mm), 14° bit can be set to cut the proper groove at ⅜" (10mm) deep, and MICROJIG has its own bit that has the dovetail with an added profile to round over the top of the ⅜" (10mm) slot for cleaner cuts and less sanding. They sell 1½" (38mm)-long "track" screws with knobs and 10-32 threads that work very well indeed with the disk clamps discussed on pages 57-61; just make the center hole ¼" (6mm) instead of 5⁄16" (8mm).

Making Your Own Clamps

Now that we have a proper spoil board setup, we need to outfit it with some clamps to hold projects in place. There are many clamp solutions available in the marketplace, and your machine may even have come with some clamps (Fig. 1). These will hold your work but not as securely and efficiently as you may wish. They require a lot of space on your machine to work, and with larger projects, this can be an issue. For example, my machine came with only two clamps, which will simply not be enough for many projects. This section will show you how to make disk clamps and arched clamps.

Low-profile Disk Clamps

Parts being milled need to be held firmly against the table, especially to prevent them from sliding side to side or front to back. Just as with your handheld router, there will be resistance or "side pressure" while cutting. Effective **work holding** clamps workpieces while being quick and easy to use. You want to be able to position and secure clamps very quickly. I have made, bought, and used any number of clamp systems, and I keep returning to these simple disks for most of my clamping needs. They are inexpensive to make, easy to use, and low profile, so there is little risk of the head hitting them. They can also be cut into without damaging the bit or machine, which is very convenient.

These disk clamps (Fig. 2) are simply 2½" (64mm)–diameter circles of ½" (12mm)–thick plywood, each with a hole in the center and a rabbet along the edge. The rabbet hooks over the top edge of a part to hold it down, and by locating disks opposite one another, you'll prevent the part from sliding sideways as well. Programming them will only take a few minutes, and you can make new ones whenever needed.

In programming low-profile disk clamps, you will discover tricks for creating clean rabbets, as well as how to structure a program so it can be run in a specific location on the machine whenever needed. The disk clamps are the ones I use most often on my machine, and they effectively hold materials up to 1" (25mm) thick.

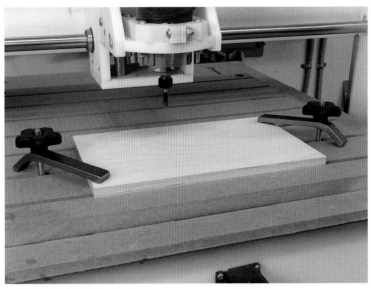

Fig. 1: An example of the factory clamps that typically come with a CNC machine.

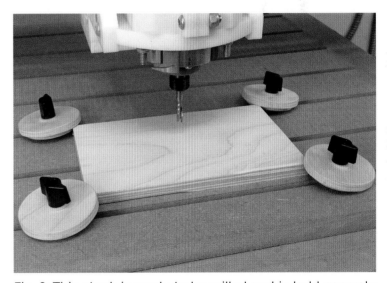

Fig. 2: This stock is ready to be milled and is held securely in place by these shop-built disk clamps.

Supplies

◇ 8" x 8" x ½" (200 x 200 x 12mm) good-quality plywood

◇ ¼" (6mm) straight bit

◇ ¼"-20 x 1½" (M6 x 1mm x 45mm) T-bolt

◇ ¼"-20 (M6 x 1mm) knob and washer

1. Start a new drawing in the CAD/CAM program (Fig. 3). Set it to 3" (75mm) square and note that the center of the work area is used as the **datum** or **start point**. You will use ¼"-20 (M6 x 1mm) T-bolts in the slots you cut in the spoil board, so the disks start with a ⁵⁄₁₆" (8mm) hole in the center. Now offset a 2½" (65mm)–diameter circle from the same center point. These define the disk. The rabbet is best defined using two additional circles, one offset ¼" (6mm) to the inside of the disk and one about ⅛" (3mm) to the outside. This overlap ensures that the rabbet will be properly defined as the disk is cut out.

2. You will want to make four disks to get your CNC set up for clamping, but I recommend programming and cutting them as single pieces rather than sets of four (Fig.4). Programming in sets means that in the future, to make even one replacement, the program will cut four parts with each run, requiring a larger workpiece. Programming for a single part allows you to cut one or a dozen without wasting time or extra material, and I will show you how singles can be easily cut from scraps too small for the set of four.

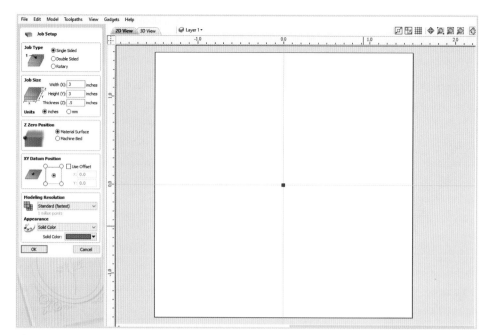

Fig. 3: Creating a new job in a CAD/CAM program. VCarve Pro is shown, but most programs will look very similar.

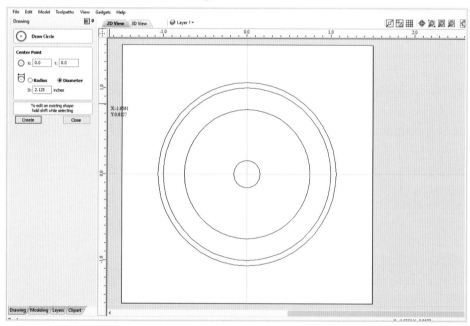

Fig. 4: The disk clamps are easily drawn as a series of concentric circles.

Tip

Whenever possible, use the same bit for all operations. Even if the run time may be a bit longer, not changing bits, resetting the Z-axis and running multiple programs will save a lot of time. With these disks, if the center hole was to be ¼" (6mm) in diameter, you would use a ³⁄₁₆" (5mm) straight bit for everything, even though this would have taken more time milling the rabbet. This is because plunging the router bit straight into the stock risks burning the stock and bit. Using a smaller bit to cut holes in a circular motion reduces friction and heat in the cutting process. This will avoid burning the bit and stock.

3. Cut the groove around the edge—the **rabbet**—using a pocket tool path (Fig. 5). Just as with milling the spoil board flat, the pocketing tool path will automatically program the tool to cut the entire selected area to the depth entered, in this case ⅛" (3mm). The pocket area must be defined by a **closed set** of vectors. The pocket can be any shape but must be fully enclosed by connected vectors. On this disk clamp, the area is enclosed by two circles, so the program knows to cut the area contained between them.

4. Note in Figs. 5 and 6 that the ramp command has been selected. This will automatically start the cutting at the surface of your part and plunge the bit gradually to the cut depth as it moves along the toolpath just as you would when using a plunge router by hand, and for the same reason—driving a bit straight down into the material like a drill bit will overheat the bit, and overheating can destroy a carbide bit in seconds. Even a very short ramp, like you are using here, will prevent burning and greatly increase bit life. The program also automatically overlaps the toolpath so the ramping does not leave unwanted material behind where the cut starts.

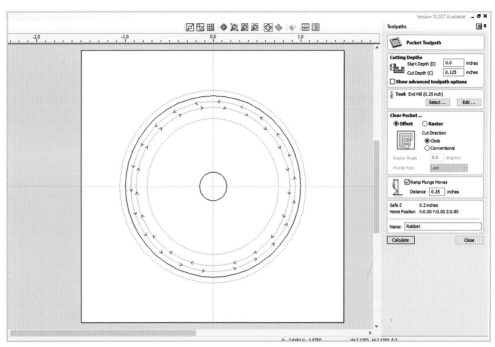

Fig. 5: The pocket toolpath is used to program the rabbet in the edge of the disk.

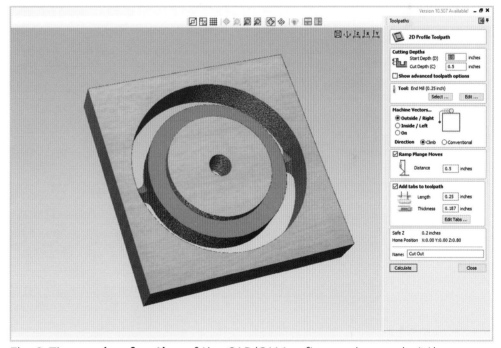

Fig. 6: The **preview function** of the CAD/CAM software shows what the project should look like after cutting. The information used to program the profile toolpath is also shown.

5. Program the center hole using the "Pocket" command. Again, we do not want to overheat the bit by drilling holes with it. So, any time we need a hole like this, use a smaller bit and the pocket command. The software will automatically **interpolate** the hole, moving the bit in a tight circle to create the proper diameter. The size of the hole can be adjusted in the program without changing bits.

Making Your Own Clamps **59**

6. Cut out the disk using a standard profile tool path with the bit offset to the outside of the vector line. Use the ramp command again and set up some **tabs** to keep the disk from moving as it gets cut out. The "Add Tabs" command allows you to define the size and location of tabs, or "bridges" that are automatically left behind to connect the part to the rest of the workpiece as their outer borders are cut away. (Fig. 6).

Even though the plywood workpiece will be clamped to the spoil board, once the disk is cut free, the clamps will not be holding the individual parts. They can get pulled into the bit, damaging the edges and even be thrown from the machine. Tabs keep the parts in place, leaving just a small area that needs to be cut to free them from the stock after milling.

Tip

Note that we have programmed all of the internal milling steps before the actual cutting out steps. With rare exceptions, this is always how milling steps should be ordered. Even when using tabs, the part may still move. With this disk project, that could mean the rabbet and/or center hole are not in the right place. It is a good habit to cut out the part from the larger workpiece last.

7. To cut the disks out, clamp a ½" (1.27cm)–thick piece of plywood to the spoil board. Remember that the center point of the disk being cut out is the start point for the program. We know the disk is 2½" (6.35cm) in diameter, so the bit can be zeroed out 3" (7.62cm) from each edge. Run the program (Fig. 7).

8. The bit will return to the start point. If you need another disk, you can **jog** the head in the X- or Y-axis, reset the zero-point, and run the program again. Because these disks are 2½" (65mm) in diameter, and the bit removes an extra ¼" (6mm) of material as it cuts them out, the bit can be jogged a known distance between runs. A distance of 3" (75mm) would be perfect here to include the overall dimension of the part, the bit diameter, and a little extra to prevent overlap. Using this technique, you can cut as many or as few parts as needed by just

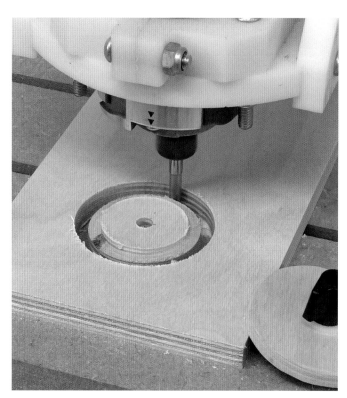

Fig. 7: A clamping disk being cut out on the CNC table.

Fig. 8: The single part program allows for easily positioning the machine to cut parts from oddly shaped leftover stock.

Fig. 9: These disk clamps can be effectively employed with stock up to 1" (2.54cm) thick. Reversing the disk allows for holding thin stock just as securely.

jogging between runs. After each jog, set a new temporary home to move all the milling operations to the new location.

9. One result of making parts on the CNC will be the unusually shaped scraps left over after running a program (Fig. 8). These scraps are known as the **web**. Often, this just needs to be cut and thrown away, but one reason for creating our program to cut out a single disk is that we may be able to fit small parts on the uncut parts of a web. This saves money and resources by getting as much out of them as possible. Because the stock is immobilized during machining, there is no need to provide a straight, clean reference edge as is often necessary for machining with other woodworking tools.

10. Install ¼"-20 (M6 x 1mm) T-bolts with a knob and washer on the disks. Keep the T-bolt as short as possible so the machine head can pass over without hitting it. On my CNC, this clearance is about 3" (75mm) with the spoil board in place. For holding materials thicker than ⅛" (3mm), face the rabbet down to catch the top edge of the stock. Finger-tighten the knob and test the hold by trying to move the stock once all clamps are positioned. You will likely be surprised by how little pressure is required.

11. The ¼" (6mm)–wide lip on the rabbet means that the clamp will only extend up to ¼" (6.35mm) onto the stock. This reduces the clamping space needed around the perimeter of the part being made, so your blanks can be smaller with less waste. If you make a mistake, and the bit does happen to run over the edge of the disk, no real harm will be done. For very thin materials, the disk can be flipped over and the flat face used to trap the stock to the spoil board (Fig. 9).

Feeds and Speeds

Your CAD/CAM software will come with some basic tools in the tool database (Fig. A), and these will have default settings that are most likely NOT suitable to your machine (Fig. B). Speed, or **spindle speed**, refers to the RPM that the bit is set to spin at, and **feed rate** is how fast the CNC head will move during the cut. On machines using routers, the RPM is set using the router speed switch and not within the program. Machines using spindles may be able to control the RPM in the program or may use some other method. Consult your manual to be sure. The feed rate is set in the program for all machines.

The feed and speed are important because they control friction. As discussed in "Bit and Collet Maintenance" (page 43) carbide bits are rarely dulled through use alone; it is usually overheating that blunts the cutting edge first. High RPM and slow feed result in cutting edges spending too much time rubbing the wood in the same area, generating friction and heat. Feeding faster and slowing the RPM both reduce friction. Ideally, the bit should run at the lowest possible RPM and the highest possible feed rate. So, what are the optimal feed and speed? They vary by bit, by material, and by machine. For example, my machine is fairly lightweight and not very rigid, so I typically run a ¼" (6mm) bit at around 16,000 RPM, feed it at 40" per minute (16mm per second), and cut no deeper than ¼" (6mm) per pass. But with an industrial machine, the same bit can cut all the way through ¾" (18mm) plywood at 300" per minute (125mm per second), spinning at 12,000 RPM.

So, how do you decide where to start with your machine? Well, beginning with depth of cut, my rule of thumb for handheld routers is that the pass depth should be the same as the bit's diameter or a little less. Because benchtop CNC routers are in the same power range as handheld routers (1 to 3 horsepower), this works as a place to start. I have also found that 50" (127cm) per minute is a safe feed rate to begin programming at until you figure out the parameters of your machine. By listening to how hard the machine is working while it is making the cut, you can tell if you need to slow down or if you can try a faster feed rate next time. If the router RPM drops a lot as each cut starts, the feed rate is probably too fast or the RPM is too low. If the bit sounds like it's chattering more than making one smooth noise, the feed rate is likely too high or the RPM too low or both. Most of us can hear when a motor is straining. With very little experience, you will be able to tell if the bit is being worked too hard or if you can increase the feed rate just by listening. Your ears are your best guide here.

Most controllers allow for overriding the feed rate during a program run but not the RPM. So if you hear the bit struggling with the cut, you can turn down the speed (which has the same effect as increasing the RPM). This override allows you to try faster rates without a lot of risk. Finally, just as with handheld routers, when you're setting bit speed on your CNC, the bigger the bit, the slower it should turn. This is because as cutting diameters increase, the cutting edges must travel farther during each revolution, so the "tip speed" is higher. A ¼" (6mm) diameter bit should be set to around 14,000 to 16,000 RPM, and a 2" (50mm) diameter bit should be at 10,000 to 12,000 RPM.

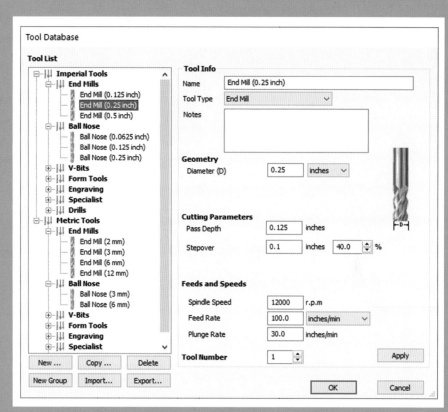

Fig. A: A standard ¼" (6mm) straight bit as set up in the default tool database that comes with a CAD/CAM program.

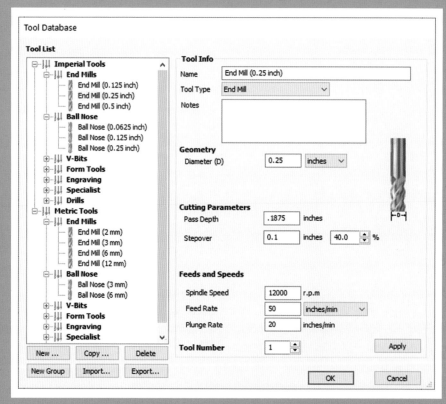

Fig. B: The default setting for tools in the database can easily be reset to fit the tools and machine at hand.

Fig. 1: This arched clamp design is very effective for holding thicker stock on the CNC machine.

Arched Clamps

For materials thicker than 1" (25mm) and other unusual clamping needs, I use arched clamps (Fig. 1). They're similar to the factory clamps that come with some machines except they can be made on the CNC. I had been making these on the band saw for some years, but the CNC is perfect for making your own in a few minutes. These arched clamps are each made from four parts that are drawn and milled to self-locate during assembly.

While drawing parts can certainly be done in CAD/CAM programs—and the drawings in this book so far have been done within the VCarve Pro software from Vectric—one of the advantages to this kind of software is that you do not need to draw every part yourself (Fig. 2). Drawings can be imported from a number of different sources such as design programs like AutoCAD and graphic programs such as Adobe Illustrator. You can simply import the DXF file of these arched clamp parts from *www. foxchapelpublishing.com/ foxchapel/cnc-machining* and go almost directly to the toolpath steps.

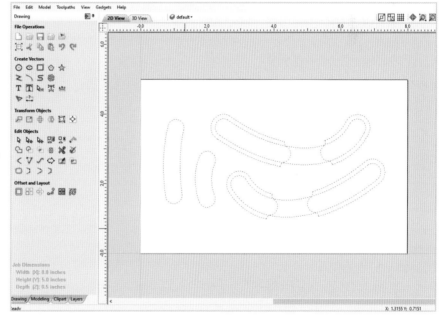

Fig. 2: There's no need to draw everything from scratch when existing files are available to import.

Supplies

◇ 8" x 5" x ½" (200 x 125 x 12mm) good-quality plywood

◇ ¼" (6mm) straight bit

◇ ¼"-20 x 2" (M6 x 1mm x 50mm) T-bolt

◇ ¼"-20 (M6 x 1mm) knob and washer

1. Start a new file and set up the job size as described in your CAD/CAM software instructions. If you are not sure of the size of stock needed, don't worry; you can always reset it after importing the file. Files can usually be imported from within the file menu or an icon in the toolbar. The DXF files are imported as vectors, so they are recognized and used in the same way as drawings created in the software itself.

2. While to the naked eye the arches look like fully connected drawings, the computer detects even the smallest gap or overlap between the end points of items. Depending on how they were created, the end points of the various elements that make up a drawing may not be connected and recognized as the objects they are intended to be. This is easy to fix, as most software can automatically interpret the design, and there are also manual tools for doing so (see Fig. 3). In this case, there should be no open vectors. All details should be closed vectors, meaning that each visible shape should be a single drawing element (see Vector Sidebar, page 66).

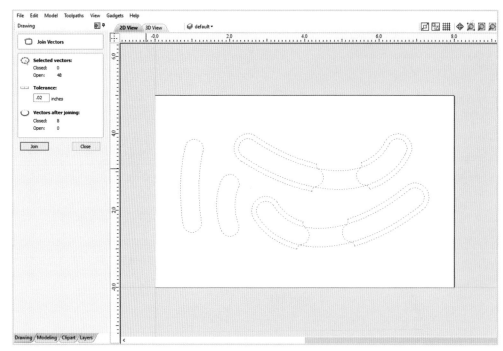

Fig. 3: Imported files may not always be properly set up for CNC machining, but the CAD/CAM software can help fix them. Here, 48 elements have been automatically joined into eight as needed: The two arched clamps, the two spacers, and the pockets where the spacers will be attached to the arches.

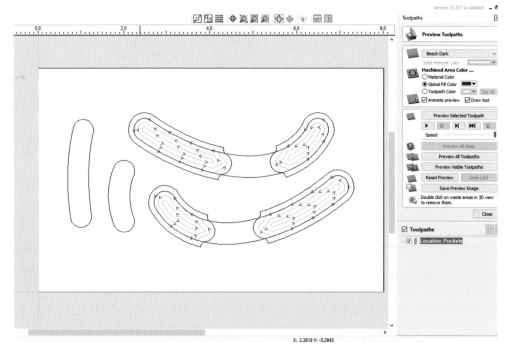

Fig. 4: Using the CNC to make the pockets, shown here by the arrows, means you can accurately set the width of the slot and use the pockets to locate parts during assembly.

There is a setting that tells the software to connect any two endpoints that are within a set tolerance. For this DXF file, 0.01" (0.3mm) was not enough to connect everything, but 0.02" (0.5mm) is. You can set this tolerance to be as large as you wish, but at some point, unrelated objects begin connecting, so be careful.

3. The DXF has the four parts that make up one arched clamp: two sides and two spacers that will be glued between the sides. There are also two pockets being cut into each of the sides to locate the spacers when assembled (Fig. 4). Similar items like these can be individually selected and then grouped together in the software.

When assigning toolpaths, an entire group can be selected by clicking on any item in the group. This saves time and eliminates missing details. It is important to note, however, that each group must be made only of items that will be milled the same way, including the depth of cut. Select all four pockets as one group, then group the sides and the spacers together as a second group. Even though they are different parts, they will all be cut out using the same toolpath settings.

4. The pockets are set to cut at ⅛" (3mm) deep. Because the spacers are ½" (12mm) thick, this will leave a ¼" (6mm)–wide slot in the center of the clamp for the bolt to pass through once the clamps are assembled. If your plywood is thinner than ½" (12mm)—or if you want to use ⁵⁄₁₆" (8mm) hardware— you can adjust the pocket depth accordingly, setting it to ¹⁄₁₆" (1.5mm) deep. The real point of the pockets is to accurately locate the spacers when gluing up the clamps.

Note in Fig. 4 that the pockets are drawn to overlap the edges of the clamp sides. This serves two purposes. First, the pockets are clearly defined as a separate feature from the profile of the side. Overlapping lines can cause confusion about which lines make up which closed vectors. Second, the bit will cut well over the edge of the clamp side, so there is no risk of material being left behind that needs to be sanded off later.

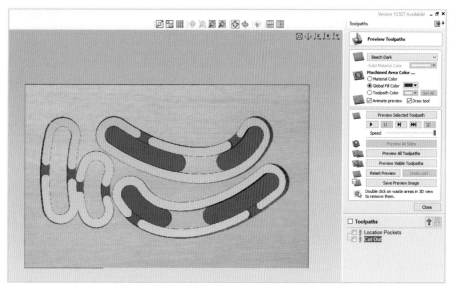

Fig. 5: This tool path preview clearly shows that the through cuts (in blue) overlap with tabs (brown) that are meant to be connected to the leftover stock (tan).

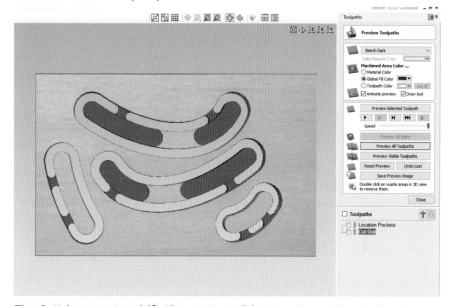

Fig. 6: It is easy to shift the part positions and use the toolpath preview to verify the program before cutting.

5. The next step is to select and toolpath the sides and spacers. These will use the same ¼" (6mm) straight bit, set to cut all the way through and offset to cut outside the vectors. Tabs have been added to keep the parts from being thrown and damaged as they are cut out. In doing so, however, the simulation of this program shows a problem. The tabs located between the spacers will cut into each other (Fig. 5). The tabs can be moved, and aligning these two together is one solution, but it is simpler to just move the parts around.

Open and Closed Vectors

Our eyes see the drawings as part outlines, pockets to be cut, holes to be drilled, and even images or text to be carved. The software only recognizes individual line and curve segments. The curved segments that make up the shape of these clamp pieces need to be joined together at the connecting points so that the software treats them as complete components. These are known as "closed vectors". Lines and arcs can be used without being connected. The lines drawn for the grooves in the spoil board project (see pages 48-55) are "open vectors". The computer treats open and closed vectors differently. Closed vectors can be cut along the outside of the lines to make a physical part, cut along the inside of the lines to make a hole of the vector shape, or as a pocket where everything inside the closed vectors is cut away to the depth you select when programming.

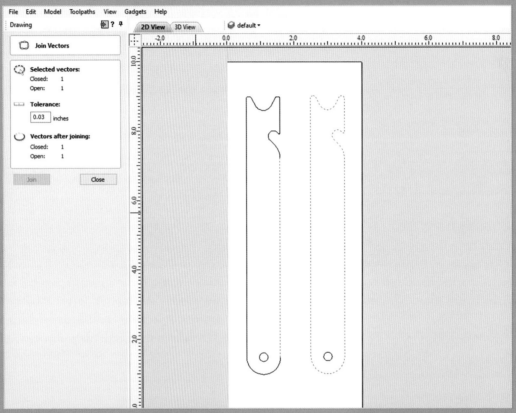

These two shapes are similar, but the one on the left is an open vector and cannot be programmed using a pocket tool path as the right-hand image is. You can usually see a vector is open when all the elements do not select as one.

Open vectors are lines and/or curves that are not connected end to end. They do not totally enclose a space. This means that they cannot be used with pocketing tool paths, since the computer does not have a fully defined space for the pocket. The software also has no clear definition of the "inside" or "outside" of an open vector, or even a set of vectors that are mostly closed, so these "inside" and "outside" will be based on the start point and direction of the vectors as drawn. How the software chooses what is "inside" or "outside" will not always be obvious with open vectors, so it can lead to cuts being made on a side that is unintended. Typically, open vectors are tool pathed with the bit cutting on the line, as in the spoil board project.

Text and carving often include both open and closed vectors and are treated differently. This is why there is a special "V Carving" tool path that is distinct from the other more basic tool paths that can be chosen.

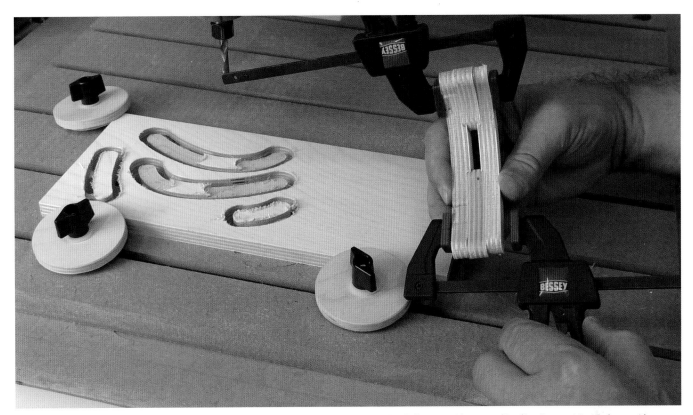

Fig. 7: This photo shows one arched clamp being assembled while another waits its turn. Note how the spacers fit in the pockets as designed.

In many CAD/CAM programs, items can be moved, rotated, and even resized, and existing toolpaths can be simply recalculated to align with the new drawing (Fig. 6). Even if you have to edit the toolpath data manually, it takes just a few extra seconds. Now none of the tabs interfere with each other, and of all the parts have tabs properly attaching them to the waste. Tabs can often be added automatically in the software, but placing them manually allows for making sure that they will work properly.

In the drawing for this project, any tabs between the two sides would be connected to a fragile section of waste that may not be solid enough to prevent movement. The tabs are located where it will be easy to sand the parts—not on the rounded ends but all in the area that will need to be sanded smooth after assembly.

6. To get the machine set up to mill the clamps, zero the Z-axis to the top of the plywood stock. The depth of the pocket is critical to control the width of the slot between the sides, so setting zero from the top of the material sets the pocket depth accurately, and any variation will be in the leftover material.

7. Both the pocket and cut out tool paths use the same bit, so there is only one program to run, and each run makes a complete set of clamp parts. The spacers get glued into the pockets on the sides, and everything is clamped until the glue dries (Fig. 7). The edges can then be sanded smooth. By cutting and assembling the parts this way, we end up with a curved clamp, nearly 1¼" (32mm) thick, with a slot running through it perpendicular to the curve of the body. You can make this clamp by hand, but not in the few minutes it takes on the CNC.

These clamps require a longer T-bolt and will be higher profile than the disc clamps we made earlier, but for thicker stock or if you need a clamp to hold a piece of odd-shaped scrap well in past an edge, these arched clamps are good to have on hand and cost almost nothing. They are also highly useful for holding parts on table saw or router table sleds.

Making Your Own Clamps 67

CHAPTER 3

PRACTICAL SHOP PROJECTS

This chapter is focused on using your CNC to make jigs, fixtures, and other items that help your more traditional woodworking better. Too often, safety items go unused because they are not at hand or are worn out. Being able to rapidly make a new push stick, throat plate, or safety shield whenever needed will reduce the risk of accidents and improve the quality of your work. This is an area where the CNC can really help improve your shop.

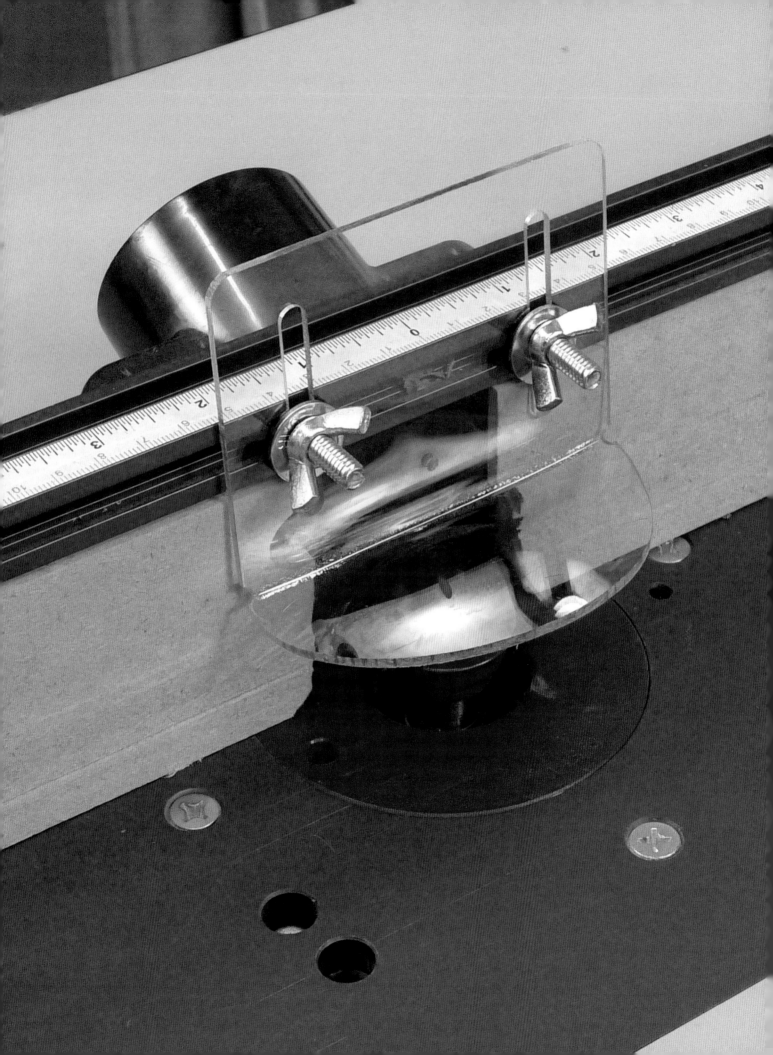

Push Sticks

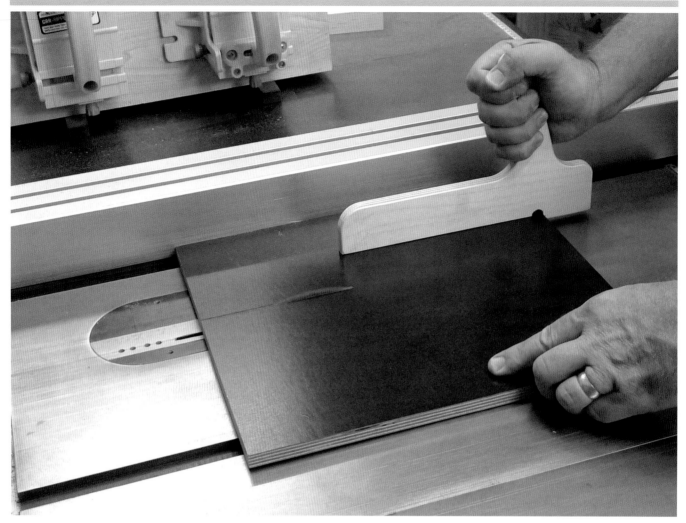

Fig. 1: A typical push stick being used to cut narrow parts on the table saw more safely.

I prefer to use a "shoe"-type push stick to help make narrow cuts more successfully and safely at the table saw. I have been making my own push sticks for many years (Fig. 1). I used to draw the shape on some plywood, cut it on the band saw, sand the edges smooth and then round them over at the router table. Now, I use my CNC to cut out the push stick instead (Fig. 11). I only need to round over the handle edges at the router table before using it. Best of all, since a CNC will replicate parts exactly, every push stick I make is the same instead of each being slightly different.

Supplies

◇ 11" x 7" x ½" (28 x 17.8 x 1.3cm) good-quality plywood or plastic

◇ ¼" (0.64cm) straight bit

1. The push stick is a good project to walk you through basic CAD drawing. It is more involved than the rectangle and circles we drew for the table mill and clamps, but it's not too hard to follow if you've never used a CAD program before. You start your CAD/CAM programming by setting up the job space. This tells the program how large the piece of stock will be. In this case, it will be a piece of ½" plywood measuring 7" x 11" (180mm x 280mm). In VCarve, these dimensions are entered in a "Job Setup" dialog box, but other programs will be similar. As you enter the numbers, a representation of the stock will appear on the screen (Fig. 2).

You also need to enter an origin point for the program to run from. You can choose to start from any corner or the center of the stock. The lower left corner is the default starting point, since with cartesian coordinates moving up and to the right are positive numbers. If I will be working with a part already cut to final size, I will start from the left corner. But most often I am cutting parts out of a larger blank like we will here, so I like to start in the center. You will see why when we set up to run the program.

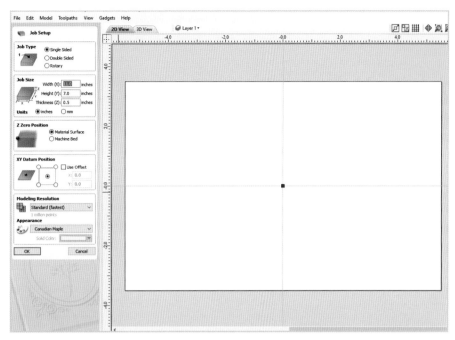

Fig. 2: Any CAD/CAM program starts with the Job Setup Dialog box where the dimensions of the workpiece and origin information are entered. This creates a virtual part for you to draw on.

2. Basic CAD drawings can be rendered by combining simple shapes, and that is what we will be doing here. Your options include drawing lines, curves, rectangles, circles and other shapes. Select the line icon and begin drawing from the lower left of the stock toward the right by clicking and dragging. You will notice that your line will be straight, not a curve. While it can be drawn at any angle, it will "snap" into place and be highlighted if you get close to horizontal or vertical. Notice also that the length of the line is shown as you drag the cursor.

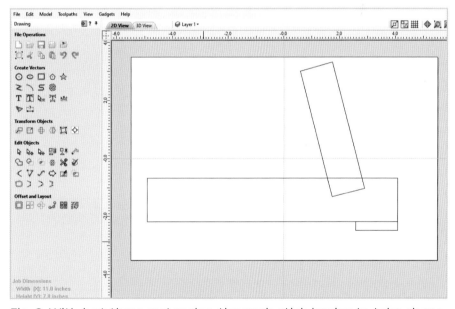

Fig. 3: With just three rectangles, the push stick begins to take shape.

Click to end the line when you have it horizontal and 9" (230mm) long. Now draw up 1.5" (35mm), back to the left and down to your start point to form a rectangle.

The push stick shape emerges when you add a second rectangle at an angle for the handle and a small third rectangle that makes the hook (Fig. 3). Double-clicking on any of these shapes highlights them and allows for moving, scaling or rotating, so you can adjust the components of your drawing

until it is what you want. The software will have a measure tool to verify dimensions when needed.

3. With the basic shape complete, cut away any sections that intersect or overlap using a trim tool. In VCarve, this changes the cursor into a small scissor image, and clicking this on any line will cut the line back to the nearest intersections. If you remove the wrong section, simply click the undo option in the edit menu like any other software program. As you remove the extra lines, VCarve will automatically join the rest, so you end up with a single push stick shape with all square corners (Fig. 4).

4. Square corners can be rounded off using a **fillet** tool (Fig. 5). I do not know why the term fillet is used, but in CAD terms it means connecting two lines in a corner by adding a radius. Type the radius you want into the dialog box and then click on any corner to round it to that radius. You can try a couple of radii and undo them until you are happy with the results. They do not need to all be the same. The body of this push stick has 1" (25mm) corners and the end of the handle is ½" (13mm). Don't worry if your design isn't centered on the workpiece. You can use the cursor to select the entire shape and either move it where you want it with the mouse, or use VCarve's "Align Objects" feature to position it precisely.

On the bottom of the push stick where the heel drops down to act as the pusher, the corner will be left with a ⅛" (3mm) radius from the ¼" (6mm) bit used to cut out the part (Fig. 6). This does not show in the drawing, but previewing the tool path will clearly show it. This corner could certainly be squared up after cutting in any number of ways, but you can fix it in the program too. Cutting a little more deeply into the corner will create a space that moves this radius away from where the stock will be, so square parts will butt up fully against the push stick's heel.

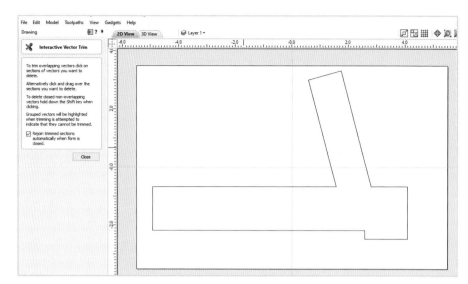

Fig. 4: Trimming the shapes where they overlap will automatically join them together into a new form.

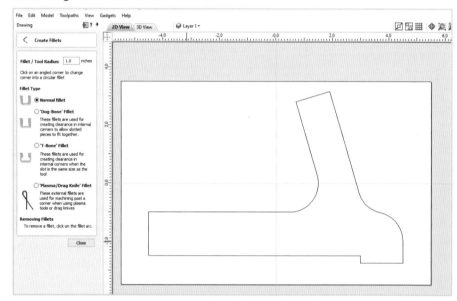

Fig. 5: Square corners can be instantly rounded using the fillet tool.

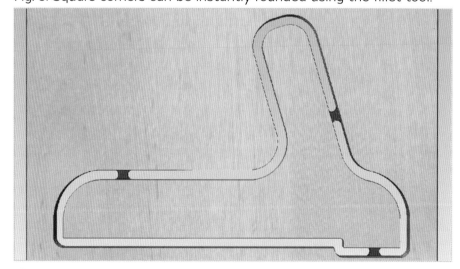

Fig. 6: Router bits cannot cut square inside corners, so the heel of the push stick will need to be squared up to push stock safely.

Preview Function

One of the most powerful tools you will have as you learn to program your CNC is the "toolpath preview" function. All CAD/CAM packages have this feature, and it provides a 3D rendering of the actual cuts your current program will make. When starting out, you will make a lot of simple programming errors. Using the preview function properly will save you a lot of wasted time and material. Barring mechanical problems, your CNC will do exactly what you tell it to do through your program. Mistakes in programming may only result in wasted stock, but in some situations, it can damage your bits and even your machine.

The toolpath preview feature will show you the part after it is milled. You can preview the result of any one toolpath, two or more or the entire program. The cutting is an animation that shows a virtual tool cutting the virtual stock in slow motion. And while your G-codes only contain toolpaths for one bit each, the preview can show all tools used. In essence, you get to "prototype" your program before cutting into any stock.

Fine details will not be obvious in the preview, so if you wanted a pocket to be cut ¼" (6.35mm) deep but programmed it to actually be ⁵⁄₁₆" (7.93mm), you may not notice the difference. But if you accidentally called for the pocket to be cut through the part, the preview will show you instantly. As you will see in the relief carving project, the preview allows you to actually see the shapes that will be cut with profile router bits, and it enables you to quickly experiment with various settings to forecast how textures will come out.

Using preview to check your work will help you spot errors and correct them at the computer. As your skills grow, the preview allows you to experiment and visualize new cut strategies without risking your machine.

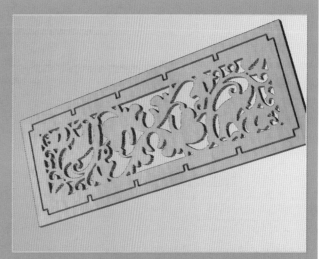

The preview function lets you look for errors in complex parts like this pierced lamp side before any material is even cut.

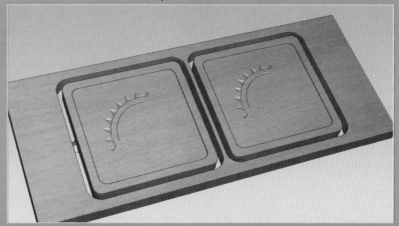

A logo is supposed to be cut into the bottom of this pocket, but the start depth was not set properly. Without checking the preview, this error would only be apparent after completing the actual cutting.

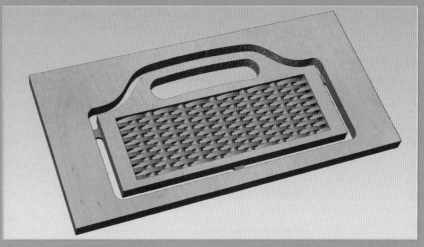

The preview function allows for trying different settings, cutting strategies and even bits to see how they will be cut before milling. The more complex the program, the more useful the toolpath preview function can be.

You remove this radius by creating a circle in the intersection that forms the corner. Make this circle at least as big as the bit so it will be cut. Use the trim tool to remove the excess. Since this sort of notching is a feature that is common in machining, many CAD/CAM packages have a fillet routine that automatically adds this extra cut into any corner you select. VCarve has a couple of options for this, and you should try them and see what the results are. You can always undo any mistakes. In this case, use the "T-Bone" fillet command with the radius set to the bit diameter of ¼" (6mm) to make the exact notch we want (Fig. 7).

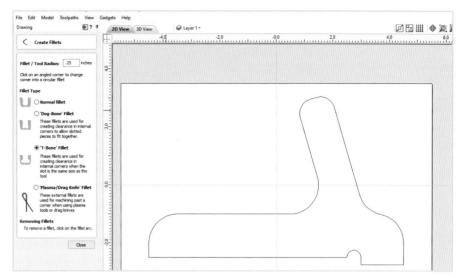

Fig. 7: Fillet commands allow for automatically adapting inside corners with one click.

5. Once your push stick drawing is complete, switch to the CAM side of the program to begin programming the actual cut (Fig. 8). Begin by selecting the outline of your push stick so it is highlighted. On the right of your screen are icons showing all the operations available to you for the CAM process. The highlighted drawing is a profile of the part, and most CAD/CAM programs refer to cutting parts out as a "profile" tool path. Selecting this icon will open a new dialog box where you will fill in the information needed to make the cut.

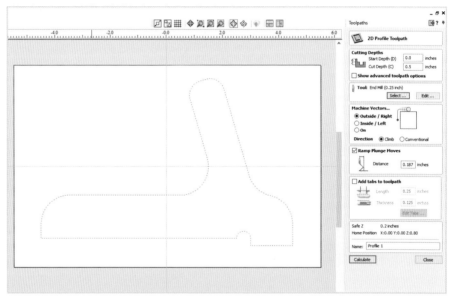

Fig. 8: The CAM part of your software is where you make decisions about tools and cutting.

There are a lot of choices here and even a box to select "advanced options". We will cover them in other projects within this book, but for now we want to create and run a simple cutout. The dialog box has the options set in a logical order from top to bottom. The first information you need to enter is the depth of cut. In this case, our push stick is being cut from ½" plywood, so enter that number first. Next, choose the bit you will use. Just as with a handheld router, the ¼" (6mm) and ½" (12mm) bits are most often used. In this case, select the ¼" (6mm) bit, since the ½" (12mm) will not work in the notch at the heel.

6. Next, the software asks how you want to machine the vectors, which means where you specifically want to cut the lines. As with any cutting tool, you need to choose if you want to cut on the line or to one side or the other. So the software offers the choice of cutting outside the shape, inside it or directly on the line. Since we want to cut out a physical part, select outside. VCarve has small images that show you which is which.

Select "Ramp Plunge Moves" next. The ramp moves the router as it plunges to prevent burning the bit as it begins the cut. The ramp distance should be set to the bit's diameter or a little less. Add tabs to prevent the part from moving as it is cut from the blank. You will cut through these tabs after milling to remove the push stick. Finally, you need to name the tool path. It may seem unnecessary with only one operation, but naming tool paths is a good habit to develop right from the start.

7. The final programming step is to create the G-code from this CAD/CAM file (Fig. 9). You will remember from "Defining the Process" (page 36) that the G-code is the file that translates all of this information you just entered into the file needed to run your machine. Select the tool path and click on the "Save Toolpath" icon. You will need to select the post processor for your machine if you have not already, and then save the program to your computer. You have just drawn and programmed a push stick from scratch. Now it is time to set up and run it.

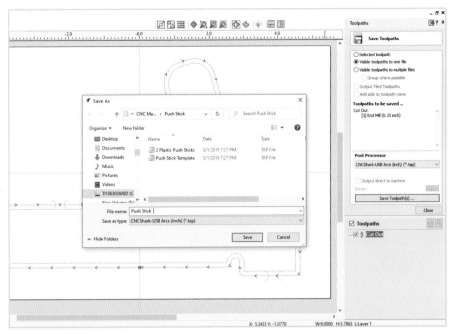

Fig. 9: Your last step before running the CNC is to output the G-Code from the CAD/CAM software. This is a distinct file created specifically to work on your model CNC.

Fig. 10: Plywood is often not exactly the thickness stated. Setting the Z-height on the proper setup block keeps the spoil board from being damaged.

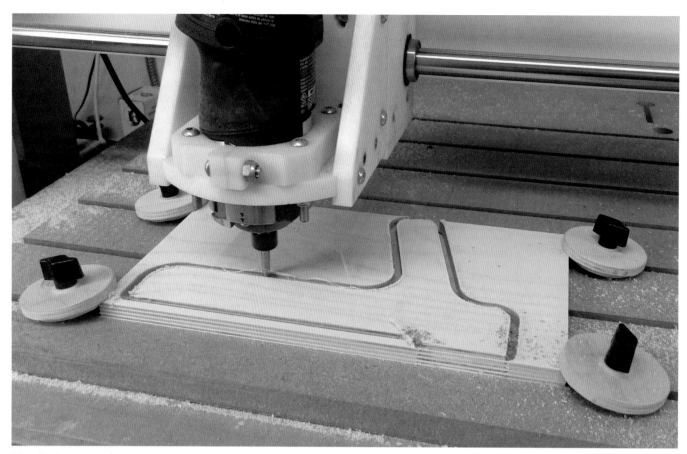

Fig. 11: Once you've written the program, you can make a new push stick quickly and easily whenever you need it.

8. In preparing to cut out this push stick, think about the material's thickness. If you have ever worked with plywood before, you'll know that ½" (12mm) plywood is rarely a full ½" (12mm) thick. But you'll often set up your tool paths for an actual ½" (12mm) part because you will often write programs long before the stock is in hand to measure.

Here's why this thickness difference matters: if you zero-out the Z-axis on top of stock that is less than ½" (12mm) thick, the bit will cut deeper into the spoil board than needed because you programmed the machine to cut exactly ½" (12.7mm) deep. You can always measure the stock and adjust the program accordingly, but the next time you want to make a push stick, the stock is likely to be some other thickness instead.

9. One option is to set the Z-axis depth using ½" (12.7mm)-thick block (Fig. 10). You can make your own or buy a set of setup blocks that are precision-milled for accuracy. Zeroing the Z-axis to a true ½" block will tell the CNC that the top of the plywood is actually ½" (12.7mm) above the bed so the bit will not cut too deeply into the spoil board. Another option is to simply program the depth of cut ahead of time to the actual thickness of the stock: 23/32" (18mm) for ¾" ply, 15/32" (12mm) for ½" and 7/32" (6mm) for ¼" plywood. This will limit the amount the spoil board is cut into but may not cut all the way through the part if the workpiece should happen to be slightly thicker.

Zero-Clearance Throat Plates

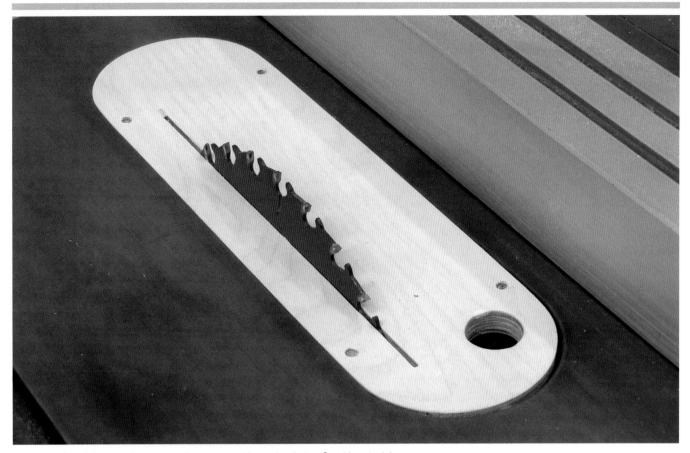

Fig. 1: A freshly made zero-clearance throat plate for the table saw.

One of the more useful upgrades for your table saw is a zero-clearance throat plate (Fig. 1), which helps to minimize tearout when cutting veneers and prevent small pieces from getting trapped between the blade and throat plate during cuts. When working with traditional tools, these inserts are a lot of work to make, which makes this is a perfect job for your CNC. All the various components can be precisely located, and you can even precut a groove on the underside for the blade before bringing it up through the throat plate. These instructions work through the process of programming and cutting an insert for a Delta UNISAW®. Your saw's table opening for the throat plate may be a different size and shape, but the steps will largely be the same.

Supplies

◇ 15" x 5" x ½" (380 x 130 x 12mm) good-quality plywood (size may vary for your saw)

◇ ³⁄₁₆" (4mm) straight bit

◇ ¼"-20 (M6 x 1mm) tap

◇ ¼"-20 x ½" (M6 x 1mm x 12mm) set screws

1. The shape of the throat plate and the location of its details can all be largely copied directly from the factory throat plate (Fig. 2). You should lay it all out upside down as seen here. Milling from the bottom allows for cutting the clearance grooves for the blade.

2. The leveling screw holes will be tapped to ¼"-20 (M6 x 1mm), so they need to be drawn in at a ⁷⁄₃₂" (5mm) diameter, the proper hole size for a ¼-20 tap.

3. Note that two boxes are drawn for blade clearance grooves. The blade opening is ⅜" (10mm) wide and 9" (230mm) long, while the dado opening is 1" (25mm) wide and 7" (180mm) long. Because we can select to toolpath only the parts we want to, we can create one CAD/CAM file that can make either a zero-clearance or dado insert as needed by only adding toolpaths to the box needed. Each clearance groove shape is a separate selectable tool-path.

4. Create tool paths for all of the features including both clearance grooves. Then output the G-code, selecting which opening you want the throat plate to have. Another option is to cut all of your inserts with both pockets present. This will work just fine on your saw because the remainder at the pockets is ¼" (6.35mm) thick, which is enough for safe table use.

5. You can make a couple of throat plates at the same time, and any of these can become either a zero-clearance insert or a dado insert as needed. Even better, the CAD/CAM file does not directly run the machine, but the G-code we will output from here does (Fig. 3). You can simply deselect the dado pocket toolpath and output the G-code as a zero-clearance insert, or you can deselect the blade pocket and output as a dado insert. Two distinct G-codes are created from one file based on which toolpaths are selected to output. This is the sort of flexibility that the CNC machine gives you in your shop.

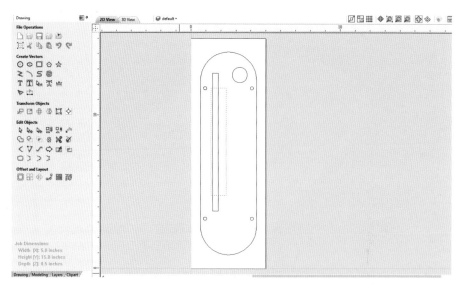

Fig. 2: The program is set up to make openings for a dado or standard throat plate.

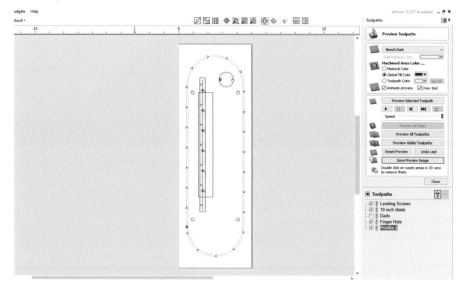

Fig. 3: G-codes can be created for both types of insert by selecting the correct toolpaths. Once programmed, either style can be made very quickly without reopening the CAD/CAD software.

Throat Plate Usefulness

There are several tools in the shop that use throat plates, and your work can benefit from replacing them regularly. I have programs written to quickly make new throat plates for my band saw, router table, and drill press table. The table saw is not the only traditional tool in your shop that can be improved using your CNC, as demonstrated in the next project for the router table.

Safety Guards

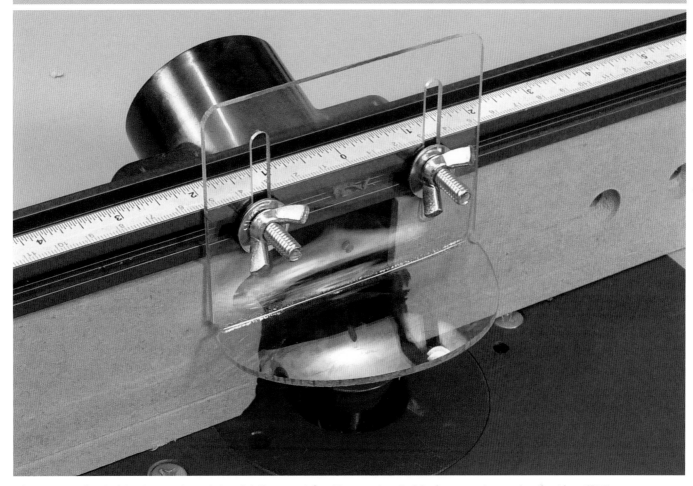

Fig. 1: An adjustable, impact-resistant bit guard for the router table is easy to make on the CNC.

My router tables get a lot of use in my shop. Typically, bits are smaller and often largely contained within the fence setup, but often I use bits and even shaper cutters that are large enough for a substantial portion to be exposed outside the fence.

In many cases, part of the cutting edge will be showing above the part being milled as well. This is when I like to install a guard to help keep my fingers away from harm. I make these guards from thin polycarbonate, which is an impact-resistant plastic. You can find it in home centers, usually in the same aisle as the glass. I tend to use polycarbonate that's 0.080" to 0.093" (2 to 3mm) thick. This bit guard is an excellent example of cutting plastics on the CNC and working with thin, flexible materials (Fig. 1).

Supplies

◇ 6" x 6" x 0.080-0.93" (150mm x 150mm x 3mm) polycarbonate

◇ ⅛" (3mm) O flute straight bit

◇ ³⁄₁₆" (4mm) straight bit (optional)

1. The bit guard requires two main features: a horizontal guard that will be positioned over the bit, and some vertical portion that can be fitted to your existing router table fence (Fig. 2). In my case, there is a T-slot at the top of the fence, allowing for the bit guard to be adjusted for height as well as positioned side to side. The first step in programming this is to create the horizontal guard part. This should have well-rounded edges to prevent catching workpieces while using the router table. After making and using many of this type of guard over the years, I just make this part a semicircle to match the diameter of the bits it will shield. A 2" (50mm) radius is more than sufficient.

2. This guard will mount to the top face of the router table fence, so it needs a way to be attached. This is done with a simple pair of ¼" (6mm)–wide slots to fit the hardware my fence uses. I made these slots 2" (50mm) long because my fence is 3" (75mm) tall. The length of the slots determines the overall length of the part. Be sure to leave space around the slots for strength. Drawing and programming your own parts allows you to tailor them to your particular machine setups. Even if you've imported the .dxf file from *www.foxchapelpublishing.com/foxchapel /cnc-machining*, you can easily alter it to fit your router table needs.

3. You will bend this bit guard 90° after cutting to form the proper shape. Because the polycarbonate is soft, it can easily be formed mechanically by simply

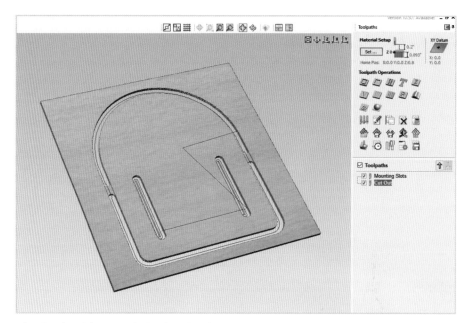

Fig. 2: The bit guard design is simple but would require several setups to make with traditional woodworking machines.

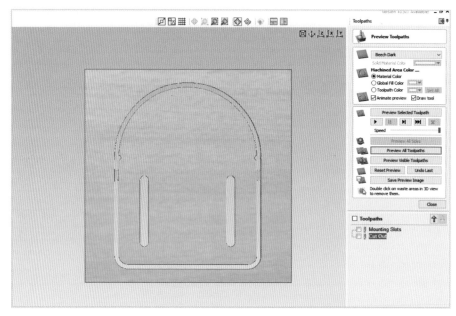

Fig. 3: The programming allows for adding highly accurate details, like these marks that indicate the bend line on the guard.

clamping it firmly and bending it, much like sheet metal. I have done this many times and have found that marking the bend line helps locate the part in the vise when bending. This can be measured and marked out, but why not simply mark it during the cutting? Your CNC makes these details easy (Fig. 3). In this case, the marks are just a pair of shallow notches on each side where the bend should be.

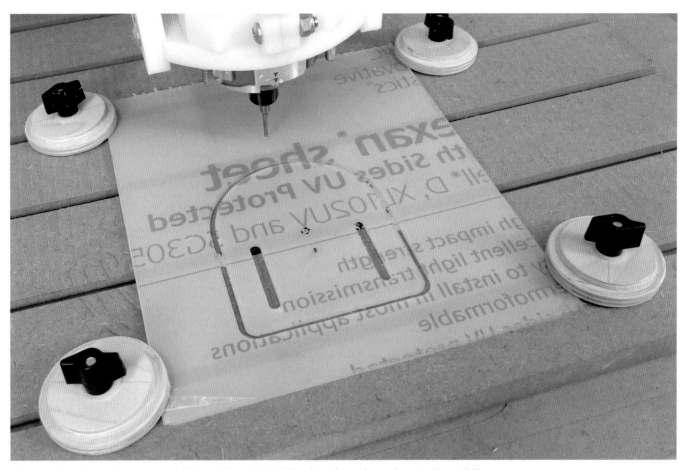

Fig. 4: Plastics cut very readily on the CNC. The key is using sharp, clean bits.

4. I chose a ⅛" (3mm) **O flute straight bit** for cutting the polycarbonate. The blade guard has to be soft and flexible to absorb impact without breaking, and polycarbonate has excellent impact resistance (Fig. 4). A standard straight bit will cut just fine, but the O flute is designed to give a superior cut and avoid heating the plastic through friction. You want to keep the RPM low and the feed rate high when cutting plastics. If the cut line gets too warm, the plastic can melt and even reweld behind the cut line. A ¼" (6mm) or 3⁄16" (4mm) bit could be used, but because the ⅛" (3mm) bit lets you make small bend markings, that is what I used. Set the feed rate a little higher than normal, 60 to100 inches per minute, and add ramps and tabs as before.

5. Once your part is cut out and the tabs are removed, clamp the round section into a vise with padded jaws and bend the top section to a 90° angle. It will spring back after the bend, but the bend line can be pounded flat using a hammer and wood block. The thin polycarbonate will not break and should bend pretty easily. You may need to overbend it a little so that it springs back to a 90° angle.

The projects presented in Chapter 3 are just a few ways your CNC machine can enhance your more traditional woodworking. We have looked at safety devices, but there is really no limit to the jigs and fixtures your CNC can create. From cutting oddly shaped templates to making precision drill guides to crafting sleds for the router table or steady rests for the lathe, your CNC can take your jigs and fixtures to another level.

CHAPTER 4

SHAPING AND LETTERING

Sign making always seems to be at the top of everyone's list of why they buy a CNC machine for their shop, and with good reason: the CNC can carve letters and graphics remarkably quickly and accurately. Plaques, signs, and similar decorative items make great gifts and are easy to market and sell for side income or even as your primary income. Graphics can be easily imported and the software pretty much automatically deals with the carving steps based on a few simple parameters that you choose. Adding text to go along with the images is easy to learn, giving you the flexibility and power to lay out your design in very sophisticated ways. There are, however, a number of practical details that you will need to know to get the best results. I have spent years figuring many of these out the hard way so I can share them now with you.

Ralph
and
Susan

★

Goshen, NH 03752
Lat: 43, 16, 42 Lon: 72, 7, 51

Carved Sign

For this project, we'll make a sign that shows a location identified on a map with latitude and longitude scribed in (see previous page). These are popular projects and easy to make with your CNC.

Supplies

◇ 14" x 9" x ¾" (350 x 220 x 19mm) wood panel

◇ 60° V-bit

◇ ³⁄₁₆" (4mm) straight bit

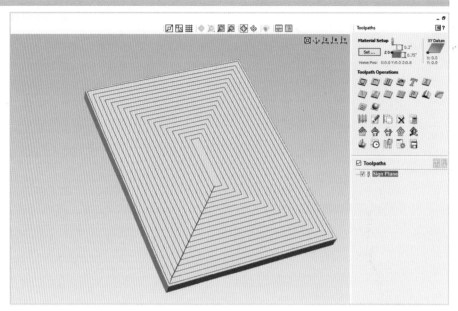

Fig. 1: The sign blank must be planed flat on the CNC to ensure that the lettering carves to the correct depth.

1. This kind of project is the perfect reason to have your spoil board planed parallel to the Z-axis as we did earlier in this book (see page 50). We will be using V-bits to carve both images and lettering for our sign. The exact depth of the V-cut is important if the items are to look as designed. Even small variations can be noticeable, and very fine lettering can vanish altogether if the stock is not perfectly flat. So, if your spoil board is all cut up and has not been planed down recently, you will want to flatten it prior to beginning your sign.

2. In addition, make sure your stock is as flat as possible and of uniform thickness. Run it through the planer if you have one. If you do not have a planer, you can use your CNC to flatten the stock (Fig. 1). Use the same type of shallow pocket setup as the table mill program we used for the spoil board (see page 50), but draw the box for the pocket toolpath to be about ½" (13mm) oversized for the stock you are flattening.

Flattening Panels and Slabs

One of the things a CNC does very well is flattening glued panels. My shop has a 6" (150mm) jointer and a 13" (330mm) planer that cover a lot of projects, but when you glue up panels to match pattern and color, the grain of the individual boards is likely to be going in different directions. Running it through your planer is very likely to tear out some of the surfaces no matter which way you feed it. Regardless, both faces of your glued panels need to be milled smooth and parallel to each other. No worries if you have a CNC: it excels at planing panels without tear out.

　　The first challenge to flattening a wood panel is that the entire top needs to be milled down, and you would cut into your shop-made clamps if you use them to hold the panel from the top. Adding some "ancient" technology to the modern, you will enlist a couple of simple wooden wedges to hold the stock (Fig. A). Clamp a pair of disks snug to the table on one side of the part and a matching pair on the other side, but leave about a ⅛" (3mm) gap between the disk and the part. Now you can tap in a couple of thin wedges to lock the part between the disks. Wedges can be quickly made on the band saw, and of course, you can cut some on the CNC as well! Keep the wedge angle shallow; too steep and it won't stay tight. If your stock is oddly shaped, like a log section or live edge slab, the disks can be positioned as needed and wedges added.

I prefer to program these operations using the center of the stock as the start point rather than one corner. Especially with odd shapes, trying to find a "corner" to reference from is difficult. If your part is largely square or rectangular, draw a box of the appropriate size, or draw a circle for odd-shaped things like tree slices. Do not even try to draw the shape of your part. Draw your pocket with plenty of extra space to ensure the tool path covers the entire part. Set up your pocket toolpath to cut ¹⁄₃₂" or ¹⁄₁₆" (1 or 1.5mm) of material at a pass. You need to mill both sides of the panel, so take as little as needed to get to flat when working the first face (Fig. B). If the entire face does not get cut after the pocket program is run, reset the Z-axis and remove another thin layer.

When you have a flat face, flip the stock over and clamp it back in place using the disks and wedges. You can set the Z-axis and take thin passes until it is fully flat when the overall thickness is not important. If thickness is critical, there is an easy process to get there. Zero your Z-axis onto a setup block and program for a specific depth

Fig. A: Wedges can be inserted between the workpiece and fixed stops to hold parts from the edges rather than the top.

Fig. B: The flattening tool path must cut a pattern beyond the edges of the panel. The wedges hold the stock securely between the clamp discs while remaining out of the tool path as the top is planed.

of cut. For example, to get a 1" (25mm)-thick slab, mill the second side down to about 1¹⁄₃₂" (26mm), set the Z-axis using a 1¹⁄₈" (28mm) stop block, and program for a ¹⁄₈" (3mm) pass depth. Since the Z is set to a known thickness, the bit may actually be cutting more or less than the ¹⁄₈" (3mm), but the cut surface will be exactly 1" (25mm) above the spoil board. Just be patient and be careful not to cut too deeply in any one pass. You are milling with a large, diameter bit just as when surfacing the spoil board (pages 50-51), so be careful not to overload the router with too deep a cut.

3. The sign will be cut out in the shape of a state—in this case, New Hampshire. Carving can be added with any custom message desired. I found a free clip art silhouette online for the state shape and downloaded it as a JPEG file. In the CAD/CAM program, set up a project with the stock at 10" wide x 14" tall (250 x 350mm) to accommodate the general shape of the image, then import the image into the project (Fig. 2).

4. In the CNC world, image files like JPEGs and bitmaps are known as **raster files**, and the lines we draw in the CAD program are **vector files**. The software cannot identify the elements needed in a raster file, so they will need to be converted into vectors, or, as in this case, the software can trace vectors onto the image (Fig. 3). Select the image to highlight it and then click the icon for tracing the bitmap. There are several options you can choose to set up how the image will be traced. The original image looks yellow, but the computer saw at least four different shades. By reducing the number of options to two (one yellow and one white), you are able to select the entire silhouette.

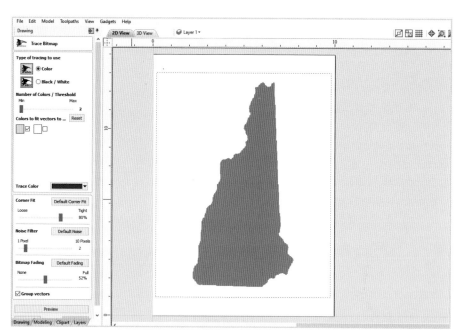

Fig. 2: Most CAD/CAM programs can automatically trace images to produce accurate vectors. This is simplified by selecting only the colors to be traced.

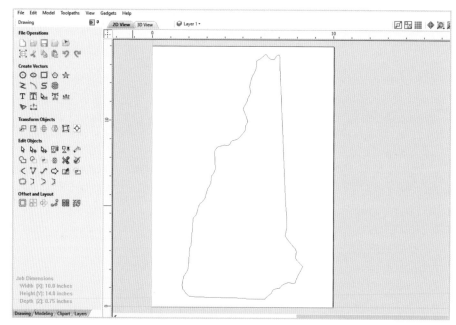

Fig. 3: The image can be deleted after tracing, leaving behind the needed vectors for cutting.

Tip

Having a sharp contrast between the image and the background will help make the trace accurate. Often I will set an image to black and white using photo editing software to provide the needed contrast. Depending on your CAD/CAM software, you will be able to make various adjustments to improve the trace, but a clean image with good contrast will always give you the best results.

5. In Fig. 4 I've added a star to this sign and text in the same font but three different sizes. If your software lets you select the actual height of the letters rather than working in font size, it can be very helpful when laying out physical signs. Your software should have come with a number of preloaded fonts, and any fonts you have saved in your computer will usually also be available. Try different fonts and styles to find what you like. The software makes it very easy to quickly try out any number of options and see the results before actually cutting.

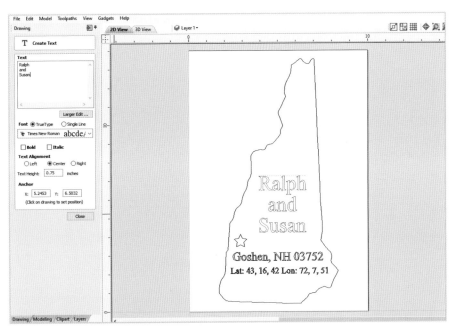

Fig. 4: The sign layout is completed with a star image and text in a couple of different sizes.

6. All CAD/CAM programs are designed to do this sort of lettering using a V-bit. Router bits cannot actually cut square inside corners, but with a V-bit, the software automatically programs the bit to rise up and out bisecting any corners, so the fine tip of the V-bit creates what looks like a hand-carved letter. It is a really fascinating example of software trickery, but it is simply based on the letters being **TrueType fonts**. The letters on the sign are each drawn as a closed vector.

When using a "V carve" type of tool path or engraving toolpath, the software will cut along the center line of the entire letter rather than following the vectors as if they were a pocket. The VCarve toolpath also does not require a depth of cut. The software measures the distance between the lines of the letter, then sets the cut deep enough that both sides of the bit touch the lines. This means that narrower letters are automatically cut less deeply and wider letters more deeply. Letters or any vectors being carved that are wider than the bit will be cut in steps at the same angle as the bit.

Fig. 5: When placed side by side, it is easy to see how the different angles change the shape of the V-bits. The most common angles are 90° and 60°.

You can specify a maximum depth of cut in the software to limit the overall carving depth. A large graphic could easily cut all the way through ½" or ¾" (12 or 19mm) stock following the angle of the bit. Setting a flat depth in a V carve tool path creates a flat-bottom like a pocket would have, but the edges of the graphic will still be angled by the V-bit.

Fig. 6: These side-by-side previews show the difference between the 90° bit and the 60° bit. The letters on the right, cut by the 60° bit, are visibly deeper than the letters cut by the 90° bit on the left.

7. Another result of the software setting depth based on the width of letters and graphics is that you can get different results by selecting different V-bits (Fig. 5). A 60° V-bit has a sharper angle than a 90° V-bit, so it must cut deeper than the 90° bit to carve any given letter. In large projects, this is not much of an issue, but with finely detailed graphics or lettering, it makes a difference. In Fig. 6 above, I have placed images of the same vectors side by side. The left sign was programmed using the 90° bit, and the right uses the 60° bit. Notice how the letters in the right image are deeper, and the smaller letters create a more noticeable effect.

8. You will likely need to sand and finish the sign after the carving is completed (Fig. 7). Often you will need to paint the letters and sand the face to remove overspray. As the surface is sanded, the depth of all letters is reduced. Be careful: this will have little effect on the larger letters, but fine details vanish quickly with surprisingly little sanding. Also, if your board was not completely flat, the depth of the letters will vary across the panel. Even small variations can distort letters or make them vanish completely. This is one of the reasons for planing your spoil board flat prior to lettering.

Fig. 7: The sign after the CNC work is completed. It only needs to be separated from the excess by cutting the tabs and lightly sanded.

9. With all of the lettering programmed, the sign can be cut out. This requires a second G-code program because a straight bit is needed. If your sign outline is a simple shape, you can use whichever straight bit you wish, but in this case, the border will require a smaller-diameter bit to show all the detail. A larger diameter bit will not be able to fit into the smaller details, especially at the top and the lower right corner. In the end, I chose a 60° V-bit to cut the lettering and a ³⁄₁₆" (4mm) straight bit to cut the sign out; this worked quite well.

10. A sign of this style can show the wood grain and have the letters and other carvings painted. I have found that applying a couple coats of shellac or other topcoat to the surface just before carving allows for easier painting of the carvings afterward. The V-bit cuts through the topcoat, so paints or stains applied to the carvings will not bleed into the wood, and you can easily sand off the excess. The finished sign was stained first, then the letters were carved through the stain, and then the entire sign was given a clear finish. This gives the opposite effect but is very easy to do successfully.

Decorative Luminary

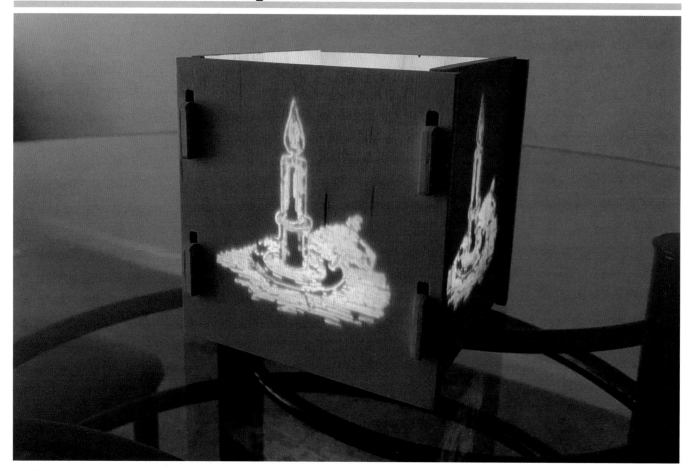

This little home décor project combines programming and machine set-up tricks to make the graphics literally glow and the parts fit together properly every time. It can be made with images suitable for Halloween, Christmas, birthdays, weddings or any event. The parts are cut with tabs and slots so once everything is milled on the CNC, they can be assembled without tools or glue and stored flat to save space when not on display.

When lit from the inside, the luminary is designed so an image shines through to the outside. This requires very precise Z-axis control to remove enough material for the light to penetrate without cutting all the way through the stock. When not backlit, the luminary looks like a simple wooden box.

Supplies

◇ 12" x 17" x ¼" (300 x 450 x 6mm) plywood

◇ 90-degree V bit

◇ ⅛" (3mm) straight bit

◇ ¼" (6mm) setup block

◇ Tea light candle

Fig. 1: Controlling the exact thickness of material left behind is an important method of programming, especially with parts that need to fit precisely, like these tabs and slots that connect the box corners.

1. Making the project's assembly tabs and slots work properly requires that the slots be sized to fit the actual material thickness of the box. Since plywood varies in thickness from sheet to sheet, this would require re-drawing the slots with each new batch of plywood. By programming the parts in the right way and setting up the CNC correctly, we can create a program that will make parts that fit using any ¼" (6mm) plywood, even if the actual thickness varies somewhat (Fig. 1).

As we did with the Push Stick project (page 70), you will zero the Z-axis off a 0.25" (6.35mm) setup block for all of the programs in this project, including the test cuts. This ensures that you control the depth of the carving and the fit of the tabs and slots using

any plywood from 0.1875" to 0.25" (3mm to 6mm) thick without needing to change anything in the G-code program. Controlling the cutting depth will also require the spoil board to be very flat. Surface yours before starting this project if it has not been done in a while, even if the CNC hasn't been used lately. An MDF or wood spoil board will absorb moisture from the air over time and become uneven as it swells.

The .dxf file for the Luminary (available at *www. foxchapelpublishing.com/foxchapel/cnc-machining*) does not include the drawing of the candle. That is a separate image labeled "Luminary Candle.jpeg" in the zip file. We'll import it in order to practice how to import a file and learn how to use the trace function.

Creating Image Vectors

2. Your first step will be to import a drawing and trace it to create the vectors needed for the machine to carve. You will be carving four copies, one for each face of the box, so we will create one, save it as a .dxf file, and use it for testing the carving program. We will then be able to import it to the Luminary program later. Begin the process by selecting a new CAD/CAM project and creating a workpiece about 4" (100mm) square to match the available space that will be on the luminaries. Import the "Luminary Candle.jpeg" image from the .zip file onto that workpiece, scaling it and moving it until it fills the space with about a ½" (13mm) boarder around the edges (Fig. 2).

Select the software's "Trace" feature to convert the drawing into the vectors needed for carving. The trace feature can convert many types of images, even photos, into vectors. It does this by tracing over the image. The trace feature works best if you use bold black-and-white line drawings. Wood block prints are an excellent source for this artwork to trace, since they use thicker and thinner lines to represent shading, which will look best when backlit. The image will also be reversed when backlit, so it may need to be flipped horizontally before tool pathing, especially if any text is included.

3. In order for the software to trace accurately, you must set several variables. Start by selecting "Black and White" as the image type, since this is what our block print is. If you import a color image, there

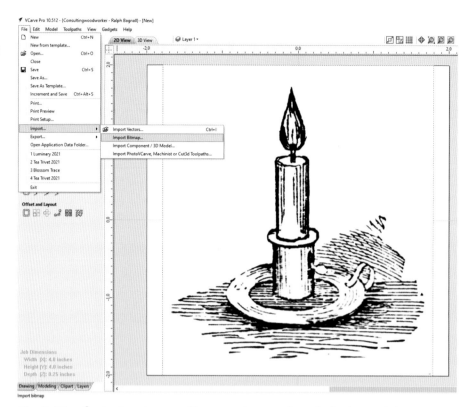

Fig. 2: The first step to creating carving vectors is to import images into the CAD/CAM software so you can trace them.

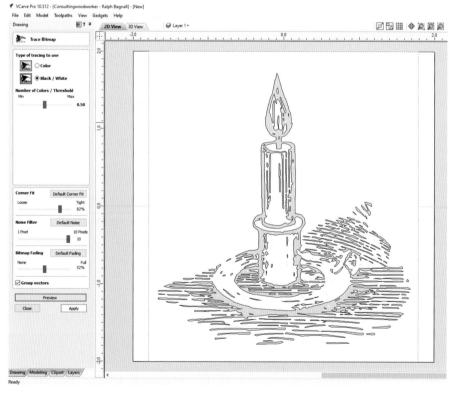

Fig. 3: The "Trace" command allows the computer to create vectors that follow the features of a raster image. These vectors can then be cut using various tool paths. There are several controls you can use to create the right vectors for your project.

will be a slide bar allowing you to choose from two to 16 colors. A set of corresponding boxes lets you further select exactly which shades to have traced. An image may be made up of several shades of the same color that form a single image element to our eye, but there is no way to carve this sort of subtle shading. So selecting the boxes for shades of a color will join them, and the computer selects them as one. (Fig. 3).

There is a "Corner Fit" function that you use to refine or round areas that come together, such as where the crossbar meets the vertical on the letter "T". The default setting is usually correct, but a slide button allows you to control this as you like. The same is true of the "Noise Filter" feature. In this case, the "noise" being filtered would be the small dots and lines that sometimes show up when images are scanned from a paper original. The slider allows for removing the unwanted pixels. The "Bitmap Fading" slider lets you make the image more or less transparent. The computer is tracing around color areas, so fading them out can change the trace, but it has almost no effect on our black-and-white candle here.

The trace feature has a preview so you can see what the vectors will actually look like. This is very helpful to understand how the program will trace your image. Experiment with the colors, corner fit, noise filter and bitmap fading and use the preview to see what settings create the best possible trace. As you experiment with importing color images, this preview will be very important. For our current project, the block print limits the number of choices, so create your trace, then select and delete the underlying imported image.

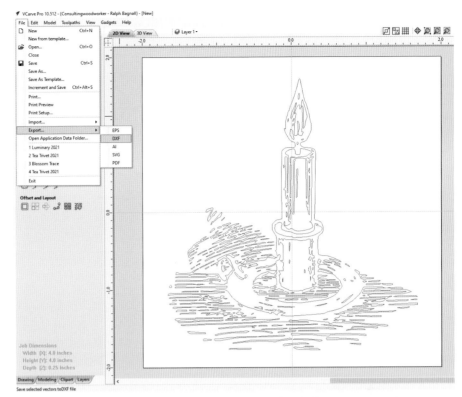

Fig. 4: Rather than convert four versions of this image to vectors, just convert one and save it as a .dxf file so you can use it on any future project.

4. This image will be carved into the back of the panels but seen from the front, so it should be reversed to appear as the original when backlit. Exit the trace function, select the entire image and click on the "Mirror" icon. This command window provides a number of ways to mirror or flip the image. Select "Flip Horizontal", which will mirror the image without moving the horizontal center line. If the image gets moved unintentionally, you can simply move it back onto the workspace using the "Align Objects" icon.

The image, now made up of vectors, can be saved as a .dxf file for use on any future project. Click on File> Export> DXF and save the file to your computer. Keeping it in the "Luminary" folder makes the most sense for the time being (Fig. 4).

Test Cuts

5. You really won't know how well the light will shine through the carving until you've machined the parts and shine a light through them. So cutting a sample will prevent wasting materials (Fig. 5). The vectors we just created from the drawing are tool pathed for cutting. This will be a profile tool path, since we need to precisely control the depth of cut for all the lines. You will carve the image and use a V-bit but not the V-carve tool path. The V-carve will set the bit depth based on the distance between lines, and that will not be nearly deep enough to show light through. Instead, we'll use the profile tool path, which will cut to the depth you assign it.

Fig. 5: The image on the left was programmed using a carving tool path, so the depth-of-cut was determined by the width of the lines. On the right is a profile tool path cut on the lines to the depth entered in the program.

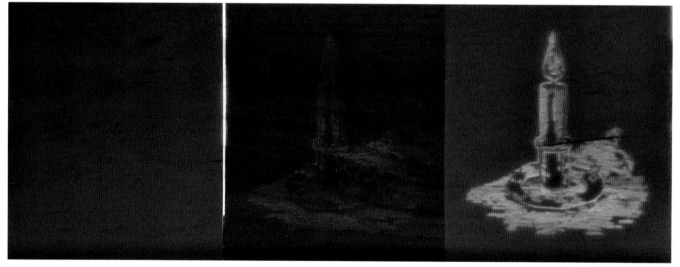

Fig. 6: These three parts show the results of the test cuts. The left part is uncut and the center part shows the first test. Strong sunlight shows through, but soft candlelight will not. The right test cut shows the desired cutting depth for the project.

6. I started by setting the Z-axis to ¼" (6.35mm) using a setup block. Then I set the cutting depth to 0.22" (5.59mm). This ensures that the stock left behind will be exactly 0.030" (0.76mm), no matter what the actual plywood thickness happens to be. After making a test cut, I discovered when holding it up to the light that 0.030" of thickness left behind was too thick to allow a tea light to shine through. So I reset the cut depth to 0.235" (5.97mm) and ran another test that seemed to allow the right amount of light to penetrate. In the photo shown here, you can clearly see how much difference it makes when changing the depth of cut by just 0.015" (0.4mm). The left piece is uncut, the center is the first try at a cutting depth of 0.22" and the right side shows the final set-up at 0.235" (Fig. 6).

Programming the Parts

7. Now that we know the ideal depth-of-cut, we can program the full project. Create a 12" x 17" work space and import the Luminary.dxf file *(www. foxchapelpublishing.com/ foxchapel/cnc-machining)* into the job. Join the open vectors into closed units as we did with the Arched Clamps (page 63). Note that there are extra boxes over the tabs and along the line of the slots. These will be used to cut pockets that leave the slots and tabs exactly 0.1875" (5mm) thick. The tabs will fit the slots properly, no matter what the actual thickness of the stock is (Fig. 7). Group these pockets together separately from the base and candle holder pockets. Those will be cut at a different depth, so we'll program them on a different toolpath.

8. Import the candle .dxf file created during the test cut and place it in the center of one box side. Copy and paste it to the center of the other three luminary box sides, too. Try to position them to match on all four sides, but don't be concerned if the image positions aren't perfect — you won't notice slight variations. While the luminary shown here has the same image on all sides, you could program yours with two or even four different images. Whichever you choose, group the images for carving using one V-carve tool path. Enter the same settings here as we found to be optimal for light penetration in the first test carving (Fig. 8).

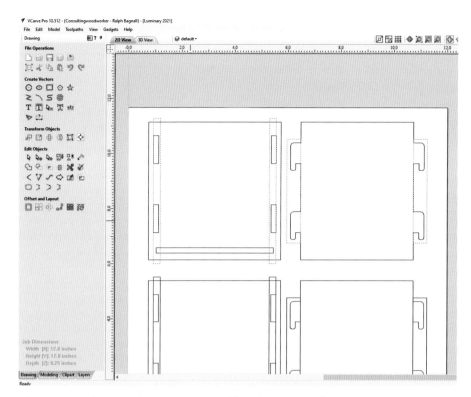

Fig. 7: Creating pockets over specific elements of a part allow them to be milled to a pre-determined thickness, no matter how thick the stock actually is.

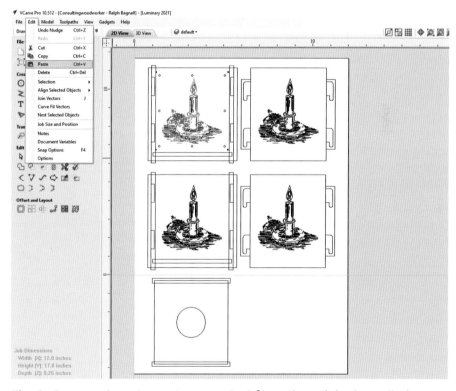

Fig. 8: Copy and paste vectors created from the original candle image into position on the sides of the luminary box.

9. Program the remaining milling using a ⅛" (3mm) straight bit. The tabs and slots will be milled to thickness by pocketing them out with a depth setting of 0.0625" (1.59mm). Set the Z-axis using a known setup block, so no matter what the actual thickness of the stock is, the tabs and slots will end up exactly 0.1875" (4.76mm) thick. This ensures that they will fit together. Note that the edges of the base are also thicknessed in this same step to make it fit in the rabbets in the sides. The tabs should end up a bit thinner than the slots they need to slide through, so keep this in mind when designing your own assemblies (Fig. 9).

10. The rabbets at the bottom of the sides accept the base of the luminary. Group these with the candle pocket in the base so they are milled together, this time cutting to 0.125" (3.18mm) deep. Next, program the slots to be cut all the way through at 0.25" (6.35mm) using a pocket tool path. Finally, all the parts are cut out using a profile tool path cutting to 0.25" (6.35mm) deep, so set that program as well. Remember to add tabs (Fig. 10).

Milling the Parts

11. Making this luminary successfully depends on controlling the depth of cut, so zero out the bit using a ¼" (6mm) setup block (Fig. 11). Choose the flattest possible plywood stock you have. Clamp the stock with the best face down onto the spoil board, and add extra clamps as needed to ensure that the stock is held flat to the table. Set the X and Y axes to zero in the lower left corner of the stock.

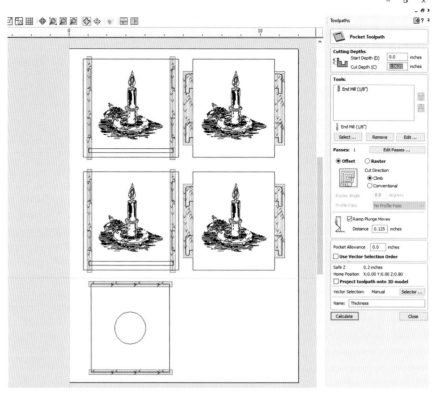

Fig. 9: You can control the precise fit of the tabs and slots by removing excess material thickness with a pocket tool path.

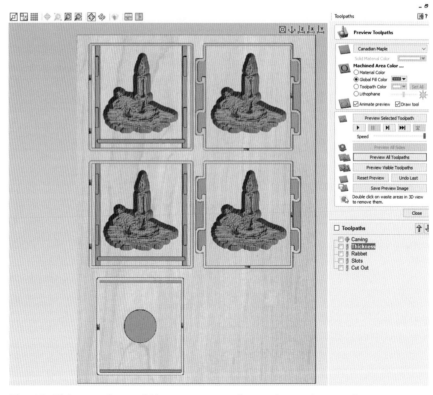

Fig. 10: This preview of the program shows how the pockets overcut the tabs and slots to ensure that they will fit together properly. Using the preview feature (see page 73) in your CAD/CAM software to verify your program will save a lot of wasted time and material.

Fig. 11: Setting the Z-axis using a setup block instead of the top of the material allows for controlling the amount of material left behind. This is more reliable method than simply using the depth of the pocket cut down from the top of the material.

12. To first carve the images, use the 90-degree V-bit. This will take some time, depending on your image (Fig. 12). These candles took nearly 15 minutes each to carve. Watch carefully to see that the plywood core does not warp and lift off the bed as it is cut away. If that happens, the bit could cut through it and spoil the image. You'll notice that the cuts do not look like typical carvings, because the result we want to see will be revealed on the other side of the stock.

Fig. 12: The four images being carved only look vaguely like the pretty candle image we started with, because we are cutting most of the way through the stock instead of carving on the surface.

Fig. 13: The luminaries are almost completely made on the CNC machine. Just a light sanding is needed before applying finish.

13. Once the V-carving is completed, mount a ⅛" (3mm) straight bit into the collet and reset the Z-height using the same setup block. The X and Y start point remains the same, so the cuts align with the carving. The thicknessing, pocketing, slotting and cutting out are all done using one tool path and bit. Note the added arched clamps in the center of the plywood in Figure 13. Add clamps wherever needed to keep the part completely flat on the spoil board. Any gap under the panel that's there initially or forms as the cutting progresses will enable the bit to cut too deeply and ruin all your careful work to control the various depths.

No Tool Assembly

14. Carefully separate the parts from the stock, sand the tabs and add a coat of finish. The luminary is ready for assembly with no other work needed. To assemble it, slide the tabs through the slots and push down to lock them (Fig. 15). Insert the pocketed edges of the base into their rabbets as the sides are assembled. Don't glue the parts together so you can disassemble these luminaries and store them flat when not in use. All that's left to do is add a tea light candle to create a warm glow that's perfect for adding ambiance to parties and get-togethers (Fig. 14).

Fig. 14: The inside of the assembled luminary shows the carving in the sides and the tea light sitting in the pocket milled in for it.

Fig. 15: Without a lit candle inside, the luminary looks like a plain box. This view also shows how well the tabs and slots fit together.

Decorative Luminary

Tea Trivet

This pretty little tea trivet project explores the use of small-diameter bits to create scroll sawn-style fretwork using your CNC; it was actually adapted from a scroll saw pattern. It is made of three parts and can be glued together permanently or left as three parts to be stored flat when not in use. It keeps hot tea pots from damaging countertops or tables, and it will add a touch of class the next time you sit down to tea with guests.

I have done some scrolling projects over the years using borrowed scroll saws, but I do not do enough scrolling to justify the cost of buying a good-quality machine. When I discovered that router bits as small as ¹⁄₁₆" (1.56mm)-diameter were available, I started experimenting to see how well they work in a CNC machine.

While there are certainly scroll saw blades that cut finer, 0.0625" (1.56mm) is a small-enough router bit diameter to be useful for a great many scroll type projects such as pierced gallery rails, fretwork shelf brackets and this tea trivet. I have also been pleasantly surprised by how durable these bits have been; through years of use I have worn out a couple but not broken any.

Supplies

◇ 7" x 14" x ¼" (175mm x 350mm x 6mm) solid-wood panel

◇ ¹⁄₁₆" (1.5mm)-diameter straight bit, similar to Freud #04-096

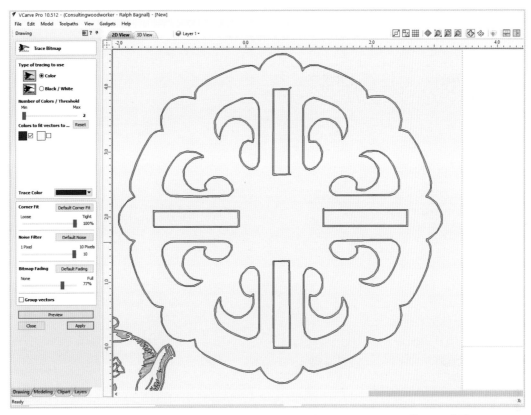

Fig. 1: Converting scanned images to rasters tends to create outlines of the features rather than a single raster line or curve. And in this scan of a trivet, the lines and curves are not straight or smooth enough to be useful.

Scroll saw patterns in books and magazines can certainly be scanned and imported into your CAD/CAM software, then converted to vectors using the "Trace" function as we do in the Luminary project (page 90). This often does not result in perfectly clean, smooth vectors depending on the resolution and quality of the image (Fig. 1). The main problem is that the computer sees the scanned lines as having thickness and so creates vectors on both sides of any line or curve, as with the TrueType fonts we use for carving the Carved Sign project (page 84). In addition, objects will likely be made up of many tiny line segments rather than smooth curves and lines. While this image looks perfectly fine, these segments result in curves being cut more slowly on the machine, since it must cut each line segment from point to point instead of one smooth curve.

I have often used a CAD program to manually trace over a pattern in order to end up with smooth lines and curves in the vectors. This can be very time-consuming if the pattern is complex, but it gets the job done. Fortunately for you, a downloadable .dxf file is available for this trivet, saving you the CAD work. It is part of the companion set for this book that you can find at *www.foxchapelpublishing.com./foxchapel/cnc-machining*.

Making the Panel

To follow along with the tutorial, the panel listed above should be glued up with the grain running along the 14" (350mm) dimension. This is typically the easiest way with the fewest glue joints; usually three pieces of wood will yield the 7" (175mm) width needed. When making thin panels, using narrow parts helps prevent warping over time. I have also made this same trivet using four pieces glued up into an 8" x 12" (200mm x 300mm) panel with the grain aligned with the 8" (200mm) dimension. After the panel is glued up, the CNC can be used to flatten and plane the panel to ¼" (6.35mm) thick if needed, as we did with the Carved Sign project (page 84).

Laying Out the Parts

1. Begin by setting up a job at 7" x 14" x ¼" (175mm x 350mm x 6.35mm) and importing the Trivet.dxf file into your CAD/CAM software (Fig. 2). Be careful not to scale it up or down; the assembly slots are sized to work with ¼" (6.35mm)-thick stock. It should automatically import at 1:1 scale. The text at the top is not needed, so select and delete it. Select the remaining vectors and use the "Join" function to convert all the open vectors to discrete features. All the interior details, including the assembly slots, are cut in the same manner, but do not group them all together. Select the four assembly slots and make them a separate group. They are sized to fit tabs on ¼" (6.35mm)-thick legs, but if your stock turns out to be a little off, you can adjust the slot width within the slot tool path without needing to redraw them. You will need to create a new G-code, but the parts do not need to be redrawn.

2. Once you have imported the .dxf file, re-orient the legs of the trivet so the grain runs along their length. This does use a bit more material, but it makes the panel glue-up easier, and the legs won't easily snap along the grain lines. Select the legs, rotate and move them into position to fit on the panel (Fig 3). The nesting feature in the software could be used, but it is good practice to do these things manually, especially when grain direction is important. (Manual nesting will also come in handy if you need to shift a part to avoid some flaw in the panel.) Once the parts are positioned, group the outside profiles as a set for cutting out.

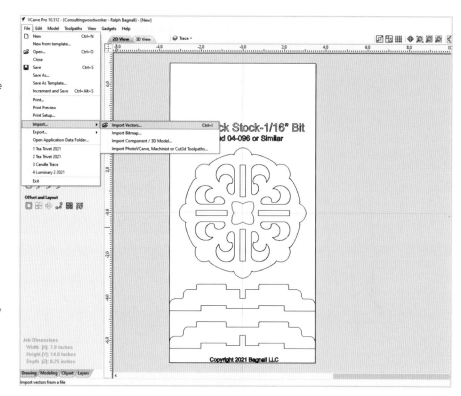

Fig. 2: Import the Trivet.dxf file directly into the CAD/CAM software using the "Import Vectors" function, and remove the descriptive text.

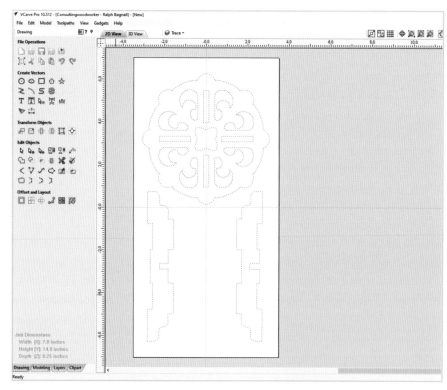

Fig. 3: The parts may need to be rearranged to match the grain direction of the glued-up panel.

Defining a New Bit

A ⅟₁₆"-dia. straight bit is not common, so it will likely need to be added to your software's tool database. Even though it's small, it's still just a straight-cutting bit, so it is easy to set up as a new bit or by copying and modifying an existing bit. Different software may have different steps, but they all work in a similar way. Define the bit by diameter first. This may automatically name the bit or you may be able to enter a name. Set the pass depth on this bit to 0.13" (3mm). This is how deep a cut the program will take with the bit. My standard cutting depth is about the diameter of the bit, but I have had excellent results with this bit cutting about ⅛" (3mm) at a pass. The "Step Over" setting refers to how much overlap will be programmed in when pocketing. Just like mowing a lawn, you want a bit of overlap. Forty percent of the bit diameter is typical, so it may be the default setting in your software already. Verify this setting in your tool database. Set the "Feed Speed" to 50 inches per minute. This may seem high for such a tiny cutter, but this bit works very well at this speed. Be sure to save the bit in the proper tool category so you can find it again later.

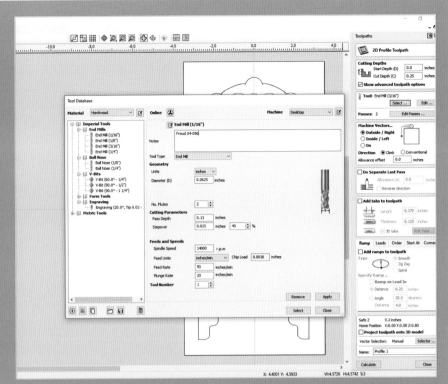

The ⅟₁₆" (1.5mm) bit may need to be added to the CAD/CAM software tool database.

Programming the Trivet

3. A straight bit is a straight bit, so filling out the tool path data for this ⅟₁₆" (1.5mm) bit is the same as for other sizes (Fig. 4). Then create the tool paths for the various features working from the inside out, starting with the decorative piercings. Set the cut depth to the full ¼" (6.35mm) thickness of the panel, and select to cut inside the vectors. Set ramps to help keep the bit cool as it plunges to the cut depth. You could add tabs to hold the small waste pieces left inside the profiles, but it is not really necessary. These pieces are too small to be much problem and often are held in place by the chips in the narrow ⅟₁₆" (1.5mm) cut path anyway (Fig. 7). When there are a lot of pierced sections, cutting and sanding tabs inside each one of the openings becomes a daunting chore.

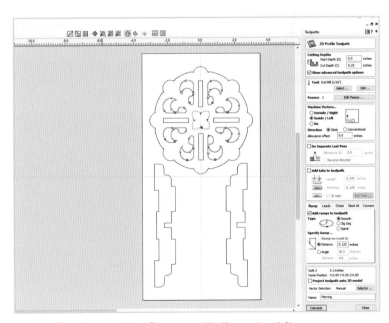

Fig. 4: The tool paths for a small-diameter bit are programmed like any other straight bit.

4. As mentioned above, program the slots to be cut out using the same tool path information as the piercings, but do this as a separate set-up so they can be more easily adjusted if needed (Fig. 5). You may end up with a panel that is thinner or thicker than planned, and having the slots separate from the piercings will save time and effort. The slots can be adjusted using the "Offset" command in the tool path rather than re-drawing and reprogramming them.

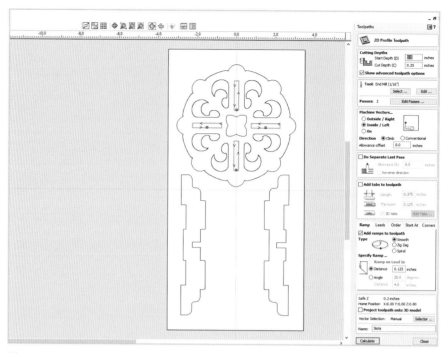

Fig. 5: Programming the assembly slots as a separate operation makes it easier to adjust them to fit if needed.

Using Leads

You will cut the parts out from the blank with the same ¹⁄₁₆" (1.5mm) bit. This saves a tool change and allows for the sharpest details on the outlines of the parts. With this project we'll use the "Leads" command. The term **"lead"** includes lead-in and lead-out. The idea is to plunge the bit into the stock outside the actual boundary of the part, allowing the bit to be fully engaged in the cut before it reaches the part cutout. This is done because all machines and bits flex at least a little when cutting. Using a lead-in removes the mark that can often be seen when the bit starts cutting on the part boundary. Leads can be straight lines, but I prefer curves and use the "Ramp" command to lower the bit to depth as it is leading in. Leads do take up extra space on the blank, and care must be taken that the leads will not cut into other parts, but they will save a lot of sanding and clean up on the edges of your parts. The lead-in and lead-out can be

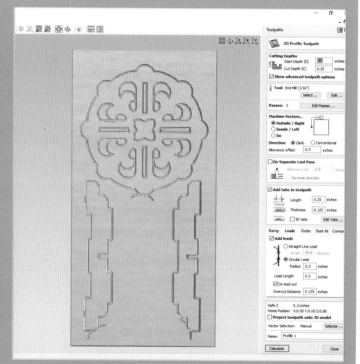

Adding a lead-in and lead-out to the cutout tool path helps remove tool marks from the parts.

programmed to overlap (Fig. 6). That will ensure the cut is consistent along the entire edge of the parts so there is less clean-up sanding to do later.

Select the "Leads" tab, click on the box to turn the leads on, then choose the options you want. You can choose straight or curved leads, a distance for the lead and how far the leads should overlap. If you click on the "Do Lead Out" button, it will use the same parameters as the lead-in. Experiment with different lead settings and use the toolpath preview function to see the results.

Cutting the Parts

5. This ¹⁄₁₆" (1.5mm)-diameter bit cuts remarkably well and is surprisingly durable. I have worn them out in use but have not broken one yet. I have found that around 18,000 RPM and a feed speed of 50 inches per minute (21mm per second) work very well in my machine. Again, you may find your machine works better with other parameters, but these are a good place to start.

6. At this point, you've now programmed the internal operations to be completed before cutting the parts out. A good safety habit you need to develop is to think about where the program will start cutting before pressing the start button. Even on a complex pattern like this trivet, the software's preview function shows how the milling will progress. Pay attention to where the machine should go first, and you will immediately see if it goes somewhere unexpected. Pause the program or activate the emergency stop as soon as possible to prevent damage to the machine or ruining your stock.

Mistakes are easy to make, especially as you learn, and the CNC will always go where you tell it to go, not where you want it to go. Often you may find that there is nothing wrong with your programmed set-up, but better to be safe than sorry. And as your skills grow, you will make fewer mistakes, but caution is always warranted around any woodworking machine. The CNC is no exception.

Fig. 6: The lead-in and lead-out overlap seamlessly, even on the corner of the trivet leg.

Fig. 7: As the tiny pierced sections of the trivet are cut free from the blank, they tend to be held in place by the chips left behind in the tool path.

Tea Trivet   **105**

Fig. 8: The entire trivet project is cut out using a single program and bit.

7. Once the internal pierced areas are completed, it's time to cut out the three parts that make up the trivet. Using the ¹⁄₁₆" (1.5mm) bit instead of switching to a larger one not only saves a tool change but also allows for much finer details to be cut into the outside profile of the parts (Fig. 8). No bit can cut fully square inside corners, but this bit is only leaving a ¹⁄₃₂" (0.75mm) radius behind, so no additional shaping or carving will be needed after the parts are cut out. The outside of the trivet top is made up of a large circle divided by a series of smaller circular segments. The ¹⁄₁₆" (1.5mm) bit will cut very clear and sharp transitions between the different radii.

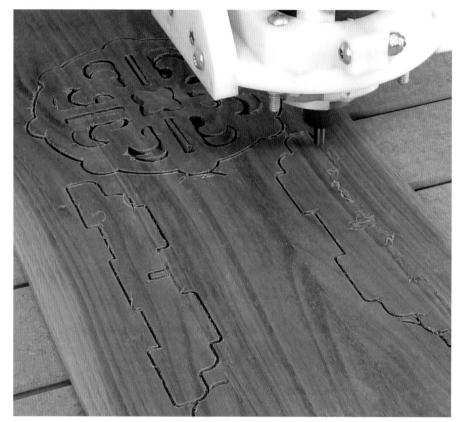

Fig. 9: Once the trivet parts are completed, they only need to be lightly sanded before finishing and assembly.

8. This is a simple example of how the CNC can be used to complete projects that used to be possible only with a scroll saw. Like many scroll saw projects, the trivet has tabs and slots for assembling it, so there's very little work to do after the parts are cut out on the CNC. They should really only need a light sanding in preparation for applying a finish (Fig. 9). Note where the lead-in and lead-out is on all three pieces before removing them from the blank. There should be almost no tool marks on the edge of the parts to indicate where the cut started. Lead-ins and lead-outs are not always necessary or practical, but when the edges of a project will be visible, it pays to set them up in your program.

Lots of scroll saw patterns are currently available as vectors, and more are being made every day. There is also an almost inexhaustible supply of printed patterns that can be converted or traced over in a CAD program for use on your CNC if you want to take the time. Your CNC can carve text or designs into the surface of your projects in ways a scroll saw simply cannot, opening up a wide range of creative possibilities for personalizing your projects (Fig. 10).

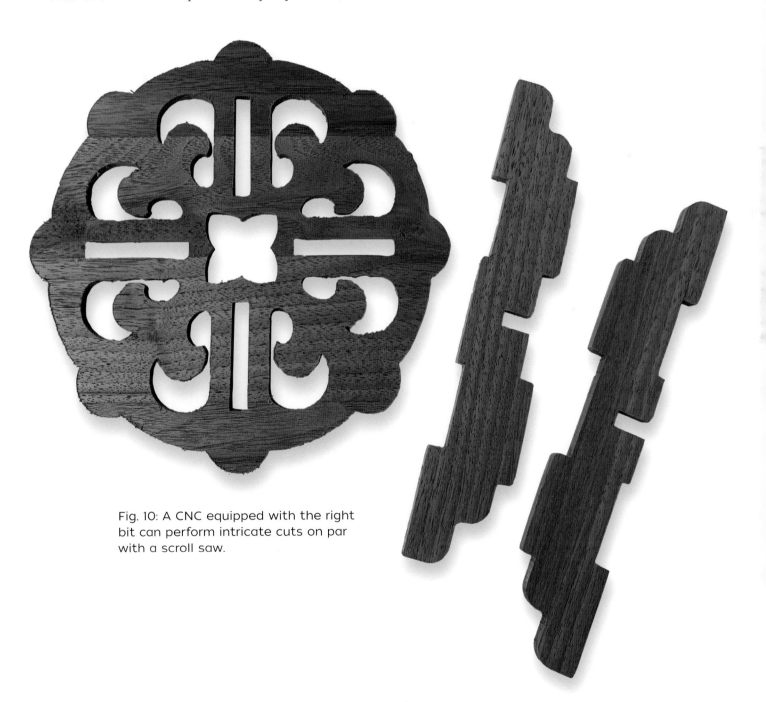

Fig. 10: A CNC equipped with the right bit can perform intricate cuts on par with a scroll saw.

CHAPTER 5

COMBINING TECHNIQUES

In this chapter, we are going to work through a couple of fun projects that use slightly more advanced techniques. We will be trying out different styles of router bits, new programming strategies, and new material choices, any of which can change how a project ends up. Once you have these options at your command, you can begin mixing and matching them to work more efficiently, do more with your machine and less work by hand after, and achieve more interesting and complex results. My intent here is both to teach you these more advanced options and inspire you to think in more creative ways about how all of these choices can improve your results.

Logo Luggage Tag

Fig. 1: The luggage tag project creates multiple parts that need to be assembled accurately.

Fig. 2: The back of the luggage tag offers a place to make it a personalized gift.

This is a practical project that requires designing multiple matching parts that need to be fitted together (Fig. 1). It features a blank surface that can be personalized with a name, signature, or logo, which makes it a great gift project, too (Fig. 2). The personalization will need to be imported and programmed for carving. The plastic window that holds the business card must be accurately matched to the wood parts in order to fit. Even though the cutting process is pretty simple with this project, all the work lies in its drawing and programming.

We have already drawn and programmed some basic designs such as the disc clamps and push sticks. But at some point, you will need to tool path designs drawn by other people in whatever CAD program they have. So we'll draw this project in a separate CAD program and import the file into the CAD/CAM program for adapting and creating the G-code. If you already use a separate CAD program, you can follow along, draw the tag and import your own files. Otherwise, you can download the Luggage Tag .dxf files from

Supplies

◇ 7" x 8" x ¼" (180 x 200 x 7mm) hardwood panel

◇ 4" x 6" x 0.080" (100mm x 150mm x 2mm) acrylic

◇ 90° or 60° V-bit

◇ ³⁄₁₆" (4mm) straight bit

◇ ⅛" (3mm) O flute bit (optional)

www.foxchapelpublishing.com/foxchapel/cnc-machining and import them into your CAD/CAM software. All of the projects in this book are available for download as .dxf files for you to make on your own machine. Find them at: *www.foxchapelpublishing.com/foxchapel/cnc-machining*.

All CAD and CAD/CAM programs have their own saving protocols. The .dxf format was created to allow files to be shared by users with different CAD programs. For example, AutoCAD saves as .dwg files, and the DeltaCAD I am using here saves in a .dc format.

Creating the Drawing

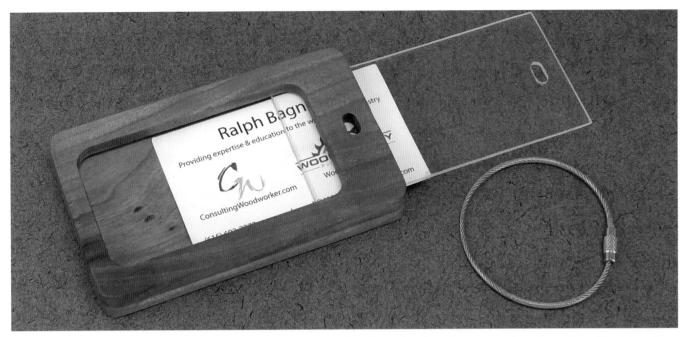

Fig. 3: The tag is made from two wood sections, a protective plastic window, and a strap for attaching it.

1. The tag shown here is designed to hold a standard business card, so all dimensions are based around that 2" x 3½" (55 x 85mm) dimension. The tag requires a hollow center to hold the card as well as an opening to make the card visible. A plastic window needs to slide inside the tag to protect the card, and it must be removable for changing cards when needed (Fig. 3). If you wish to use a different-sized card, simply adjust the initial dimensions to fit your card, and everything else should follow by adding the same offsets.

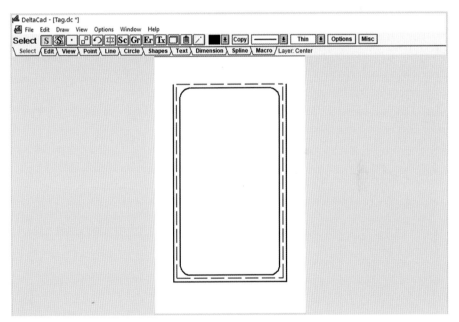

Fig. 4: All of the tag features are based upon the size of the card to be displayed as shown by the blue dotted line.

2. Three parts are required: a wood top with a pocket and opening milled in, a wood back, and a plastic window. The parts are made from rectangles of various sizes, all based on the card to be displayed. So begin by drawing a rectangle the exact size of your card, shown in Figs. 4 and 5. Everything else will be designed from these dimensions. We can use offsets from these lines to fix details. The viewing window is made just a little smaller than the card to frame it properly. Offset each of the card lines ¹⁄₁₆" (1.5mm) to the inside and add a small radius to the corners. Make the pocket that holds the plastic window larger than this opening, so the sides and bottom of the card are offset to the outside by ¹⁄₁₆" (1.5m), and the ends extended to meet each other. Leave the top open for the moment, as it will be completed as you go.

3. Sufficient space is required for gluing the top and bottom sections together beyond the viewing window, and a through hole needs to be provided at the top for the tag hanger. You can create the borders of the tag by offsetting the sides and bottom of the original card ⅜" (9.5mm) outwards and again extending the ends to meet. The wood parts will not be glued together along the upper section because the card and window need to slide in and out. Create an offset ¾" (20mm) from the top of the card to provide room for the hanger hole. Once you have drawn and connected all the outer lines, radius the corners to prevent them from catching on things in use.

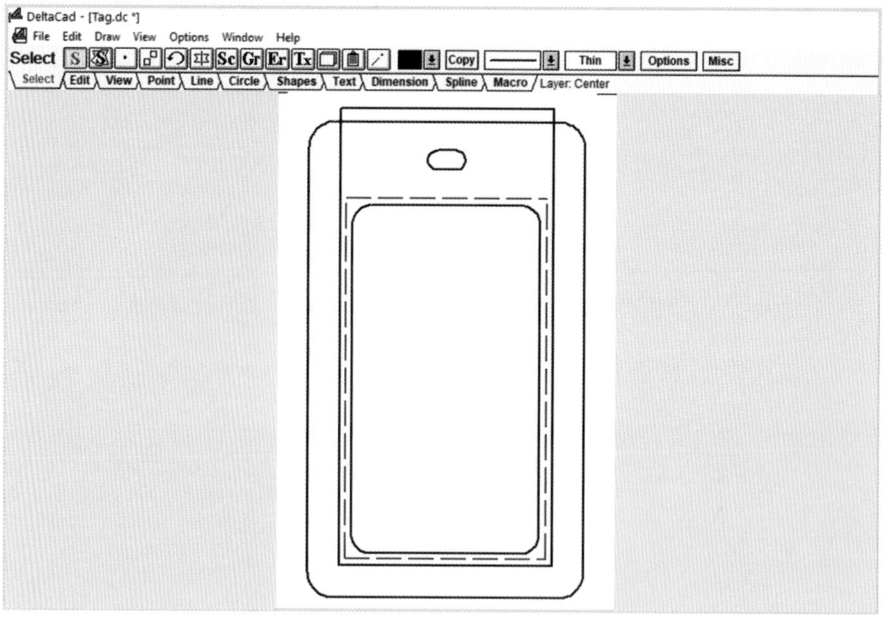

Fig. 5: Here, the top section of the tag is completed with the window pocket extending beyond the edge to create a slot.

4. Note that the pocket for the plastic window is drawn beyond the top of the tag itself (Fig. 5). If we end the pocket flush with the tag body, there will be a little arc of stock left in each corner from the round bit that you will need to trim off. By extending the pocket past the edge by at least half of the bit diameter, these extra pieces are removed during the milling.

5. Finally, you need a through-hole for the tag hanger. Draw a hole here at ⅜" long x ³⁄₁₆" wide (10 x 5mm) will work with common hangers that you can buy, while also retaining as much wood around it as possible for strength. Because the width of this slot is ³⁄₁₆" (5mm), the ³⁄₁₆" (4mm) diameter bit is used rather than a ¼" (6mm) cutter, which tends to be the default

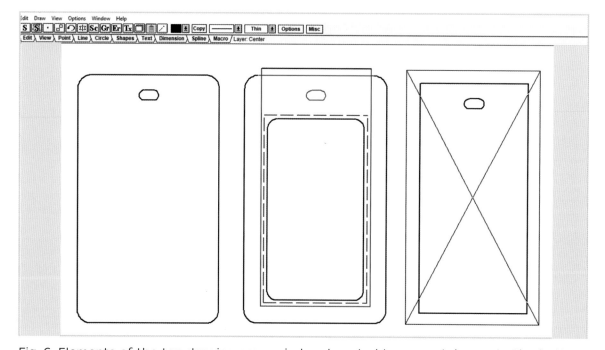

Fig. 6: Elements of the top drawing are copied and pasted to accurately create the bottom and window drawings.

tool. Milling the pocket with the smaller bit takes only a little longer and eliminates a tool change, while allowing for the smaller hanger opening. Here's another good example illustrating that bit choices do matter.

6. Select and copy the outer border and hanger hole already drawn and paste them. This is the complete bottom drawing (Fig. 6). We know that everything will line up because they were copied and pasted as a group. The personalization will be done within the CAD/CAM software by importing a graphic, so that can wait for the moment.

7. To create the plastic window, select the sides and bottom of the pocket, the hanger hole, and the upper edge of the tag itself, and paste them to the other side of the top. The top edge and hanger hole of this window piece must be the same as on the tag, but the sides and bottom need to be offset very slightly—about ⅟₃₂" (8mm) toward the inside—to ensure that the window fits in the pocket within the body. As long as you do not alter the top edge or hanger hole, everything will align when assembled because all are taken from the same drawing. Delete the blue lines that we started with; they are not to be cut.

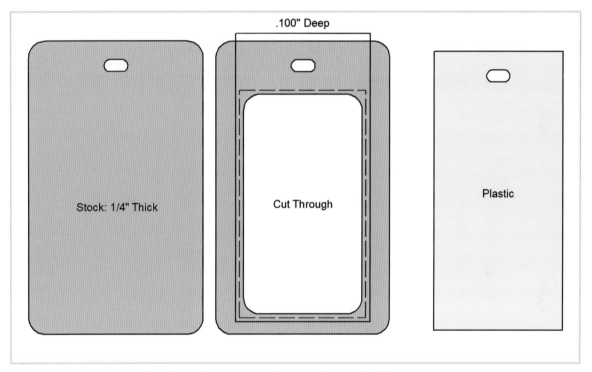

Fig. 7: A graphic showing the three parts that make up the luggage tag.

Importing the Drawing

Import these drawings into the CAD/CAM software as a .dxf file (Fig. 7). With multiple parts like these, you might like to add text notations right in your drawing to identify parts and materials. This will help you remember details if you ever decide to use the drawings again. The text can easily be deleted once the vectors are imported into the CAD/CAM software or simply left in place but not assigned any toolpath. As long as you do not select them, objects are ignored by the software. At this point, select and delete the plastic window section because it is a different material than the body parts and will be cut separately.

The VCarve software allows for opening .dxf files as existing VCarve files. It will automatically set up the project size to exactly fit the .dxf file. You can also set up the project size first and then use the import commands to bring the .dxf file in. All CAD/CAM packages allow for easy importing of .dxf files. Images and vectors can both be imported, so be sure to choose vectors.

Rasters, Vectors, Bitmaps, and DXF files

In the CNC vocabulary, image files are very different from CAD drawings, and these are known as **rasters** and **vectors**. It can get a little confusing because other terms are often used as well. In the VCarve CAD/CAM software I use, I can choose between importing vectors or importing bitmaps. In this case, the term **bitmap** represents any image file, be it BMP, JPEG, PNG, or something else. When I refer to the drawings within this book, I am generally referring to the objects drawn within the CAD section of the CAD/CAM software. These are created as vectors from the beginning and can be recognized as such by any CAD or CAD/CAM software.

Rasters are very different than vectors (Fig. A). Raster is a term that encompasses all image formats, be they hand drawings in pencil, high-resolution photographs, or graphic art rendered in an illustrator program. With few exceptions, these images simply are not recognized by the CAD/CAM software and so cannot be assigned toolpaths for cutting (Fig. B).

I am not sure why the software has an "Import Bitmap" option instead of saying "Import Rasters", but most CAD/CAM programs I have worked with use "bitmap" instead of "raster" to include all image files, even though "raster" would be more correct.

You will come to know and better understand the difference between rasters and vectors as you work with your CNC more, but, for now, remember that anything you create in your CAD/CAM software will already be in a vector format. Any sort of image file you import will be seen as a raster, even if it is referred to as a bitmap.

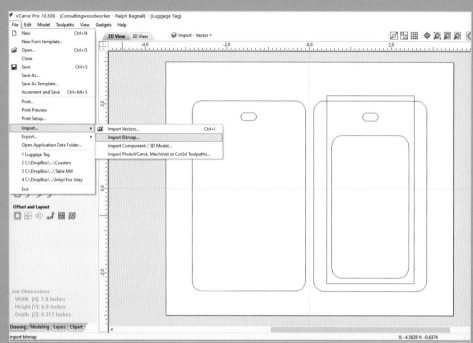

Fig. A: CAD/CAM programs handle images differently than vectors, so they must be imported separately.

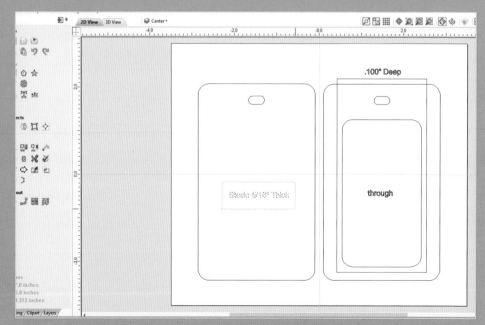

Fig. B: Labeling elements can be very valuable when drawing, but are not used in the toolpaths, so they can be deleted or simply left unselected.

8. With the luggage tag parts in your workspace, this is the time to add any lettering or graphics. If you were to add text when the part is created in the CAD program, the letters would get imported as individual objects rather than text. By adding the text now, after the .dxf file is imported. into the CAD/CAM software, it will be seen as a font and can be changed and manipulated much more easily (Fig. 8).

For this project, we are using a company logo. It is already a vector file, so it can be directly imported. The software reads the graphic as a single entity, so it can be selected, resized, rotated, and positioned easily without distortion. If you are using text you added instead of the logo, it can be manipulated in the same manner. In this project, the logo was rotated 90 degrees, resized, and moved to the center of the bottom section of the luggage tag.

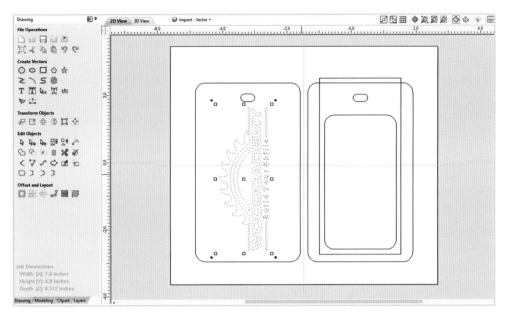

Fig. 8: The imported logo is resized and positioned for carving into the back of the tag.

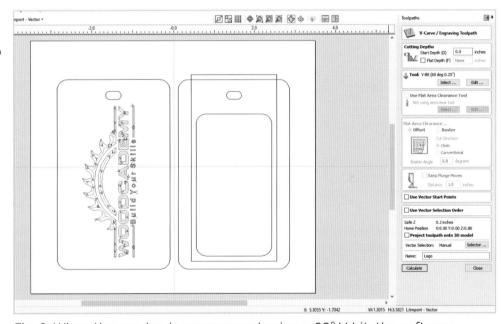

Fig. 9: When the carving is programmed using a 60° V-bit, the software automatically determines the final cut paths to create the hand-carved look.

Programming the Toolpaths

9. The carving of the logo is the only part that will be done with a V-bit (Fig. 9). Everything else in the milling process will use the ³⁄₁₆" (5mm) straight bit, so program the carving to run first to avoid more than one tool change. From the Carved Sign project (page 84) we learned that the carving toolpath works differently than other toolpaths; rather than cutting on or up to the lines, the program reads the distance between lines and sets the cut depth so that a V-bit will reach both sides. Although router bits cannot cut square corners, the carving toolpath automatically moves the pointed bit up and out at a 45° angle, bisecting the corner to create a square profile at the surface. Letters will look as if they were incised with a carver's V-gouge.

10. Another item to note in setting up the toolpaths is that the window pocket and cutout are two different toolpaths that overlap. This is not uncommon. I will typically cut the pocket first because it completely covers the window cutout. This means that the first pass of the window cut is likely to run above the bottom of the pocket or only cut in a little bit. In the Custom Coaster Set (page 126), we will look at how to fix this. For this project, the time wasted is minimal, and if we need to change the pocket depth to use a different plastic, it will change the settings for the window cut as well. For making one or two luggage tags, there is no reason to alter anything. If you should decide to make multiple tags, the effort to eliminate any wasted time will pay off.

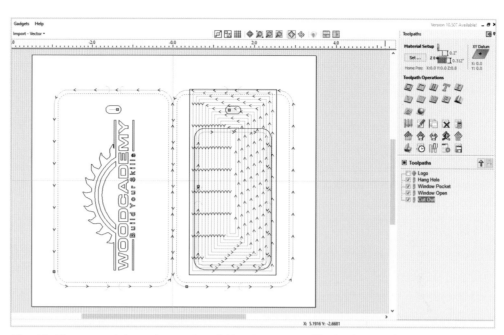

Fig. 10: The remaining elements of the tag are milled with a single straight bit. The arrows show the direction of travel that can be controlled in the CAD/CAM software.

11. Use the ³⁄₁₆" diameter straight bit to complete the rest of the milling steps for the tag (Fig. 10). The pocket is milled, hanger holes routed out, and the outside profile is cut using tabs to keep the parts in place. Use small tabs to hold the two parts as they get cut out. The carving step and the milling steps need to be output as separate G-codes because they use different tools. The programs for the wood parts of the luggage tag are ready to be run on the machine.

Applying Tabs

If parts are cut completely free from the stock during milling, they will almost certainly move as the bit removes the last piece of material. Because the bit will often be touching both sides of the kerf it is cutting, it is going to at least nick the edge of the moving part. There is a good chance that the bit will actually grab the part and fling it. This is not only dangerous, it will also damage the part, the bit, or even your CNC machine. Adding **tabs** to the cutout toolpath holds the parts in place as they are cut out (Fig. A). All modern CAD/CAM programs allow for creating tabs, and while this can be done automatically, I prefer to place them myself for a variety of reasons.

To begin with, the computer will place tabs based on the start point of the toolpath or the drawing, and this may not always be ideal. The tabs are cut to release the part from the stock and then will need to be sanded smooth. A tab on the corner of a part or on the inside of a curved part—like the arched clamps we made earlier—is much more work to sand smooth than one that is in the middle of a flat edge. Also, the computer locates tabs based only on the part being tabbed. So, when parts are tightly nested together to conserve material, the tab that should be holding part A can be cut free if the toolpath for part B is too close. Furthermore, if one edge of a part needs to be nicely finished and the other does not, you certainly want to put tabs on the unseen section.

In VCarve, it is pretty easy to set tabs manually. The tab sizes are defined, and they can be added by setting the cursor on a line and clicking or removed by clicking on an existing tab. You can also drag them along the vector line until you find a spot you like.

Designing for Plastic

12. Since the plastic window part is a different material and at a different thickness than the wood body, it must be programmed separately (Fig. 11). Set up a new 4" x 6" x ³⁄₃₂" (100 x 150 x 2.5mm) job in the CAD/CAM program, import the luggage tag DXF file again, and delete all the wood parts. The toolpath is completed in the same way as the toolpath for the wood parts. With a sharp bit, the results will be excellent. You can use the same ³⁄₁₆" (4mm) bit that cut out the wood parts, but I switched to a ⅛" (3mm) O flute bit. I have found that the O flute design cuts plastics very nicely, and I happened to have one on hand.

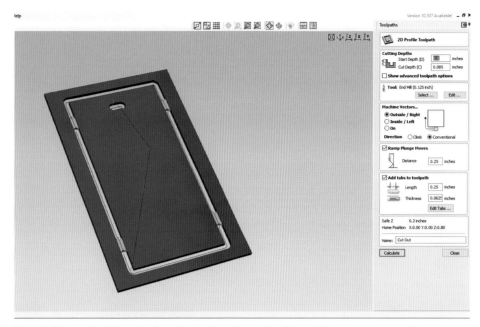

Fig. 11: The plastic window is a simple cutout and is programmed separately. CNC machines can be very effective for cutting plastics.

13. Even more so than with wood, heat is a concern when programming for plastics. It won't scorch like cherry might, but it can melt, and the melted plastic can cool and harden behind the bit, making a mess.

The toolpath and bit choices need to be made with reducing friction in mind. The O flute bit is a great choice, but any clean, sharp bit will cut common plastics well. Because friction is the cause of the heat issues, you will want to set the RPM a little less than typical for wood, and the feed rate a little faster. This seems backward, but the longer the tool spends in

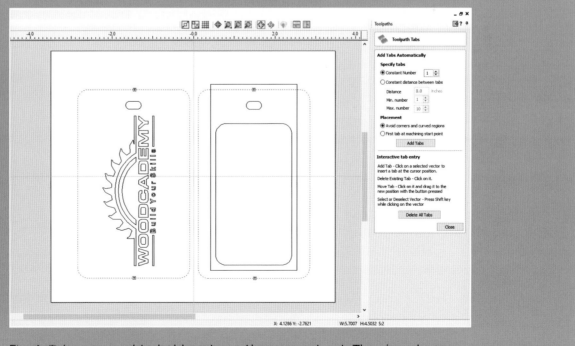

Fig. A: Tabs are used to hold parts as they are cut out. The size, shape, and locations are easily set within the toolpath dialog box.

one area, spinning and rubbing on the edges of the cut, the more friction is generated. Moving the tool along the path faster and lowering the number of revolutions both work to reduce friction. The O flute bit has a single flute (cutting edge) which produces half the friction of a standard straight bit that has two flutes.

14. Tool path the window to be cut from a piece of 0.080" (2mm)-thick acrylic or polycarbonate. These can be purchased inexpensively at any home center. While there are specialty router bits specifically made for cutting plastics, they are not required for a simple part like this. Again, use tabs to hold the part to the excess, but they have to be quite slim because the material is very thin.

Milling the Tag

15. The machine setup for milling is really no different than when making the hold-down disks and other parts we have already created. We will mill the wood parts first. Remember that projects should be cut from the inside out in case the pieces move when cut through at the end. This is still true when using multiple tools and programs. In this case, mill the carving first and the cutout second (Fig. 12).

16. Start the carving by mounting the V-bit in the machine and setting the Z-axis to the surface of the wood stock. Zero the X and Y axis to the center of the stock and start running the carving G-code that uses the V-bit. Do not worry if the head jumps around when carving graphics or letters. It is not an error if your machine skips a letter, as it will come back to it.

17. Lettering and carving like this is also why it is important that the spoil board is milled parallel with the travel of the head and your stock is a consistent thickness. Variations in the Z-dimension will show very clearly in any fine detail at the surface. The smaller the lettering or logo being used, the more likely any Z-height variations will spoil the result.

18. Mount the ³⁄₁₆" (5mm) straight bit into the router and reset the Z-axis with the new bit. Do not reset the X- or Y-axis because both programs are aligned to the same start point. This bit will do all the milling, including cutting the parts out from the larger blank (Fig. 13). While the pocket could be cut a little faster using a larger bit, it would require another program and tool change, which will certainly take longer than the extra few passes with the smaller bit. Thinking about the best tooling options can save a lot of wasted effort when using your machine. While the tabs were left to hold the parts, they can break. This is why you should set the cutout to be the last step whenever possible.

Fig. 12: The carving being cut into the luggage tag stock.

Fig. 13: The two tag components are pocketed and cut out using the same bit.

Milling the Plastic

19. The only real difference between setting up the cut for the plastic instead of the wood body is the thickness (Fig. 14). Because the plastic is less than ⅛" (3mm) thick, I flipped the clamping disks over to hold the thin stock. (That is one of the reasons the disk clamp design works so well; it is simple to make and use but very versatile.)

20. As the tool is cutting the plastic, watch the waste coming off. There should be mostly fully formed

Fig. 14: A clean, sharp bit cuts plastic easily and leaves a clean edge.

flakes of plastic coming from the cut, not dust. If there are no flakes, you can increase the feed speed or decrease the RPM, or both. Either one will reduce friction in the cut. Increasing the feed rate will also shorten the run time, so that is where I tend to adjust. Taking note of how this project cuts will tell you if your programming choices were right or if they need to be adjusted. With experience, you will build a real knowledge base that will allow you to choose the right feeds and speeds the first time.

Assembling the Luggage Tag

21. Cut the wood parts from the waste and sand the tabs smooth. Carefully glue the front and back together, and sand the seams when dry. Be sure to carefully align the pieces when clamping them

together. Doing so will reduce the sanding needed on the edges. Be careful about sanding over the carving. It is all too easy to remove fine details if your sanding is too aggressive. Apply a finish to the wood body and allow it to dry completely before you add in the card and window (Fig. 15).

22. Slide the card into place through the slot in the top first, then slip the plastic window in between the card and the viewing opening. Line up the hanger holes in the wood and plastic and insert whatever type of hanger you wish. The cable shown in Fig. 16 is much easier to attach than the silicone loops that can be bought, especially with a tag this large. In either case, longer is better to easily connect to the handles on your bags.

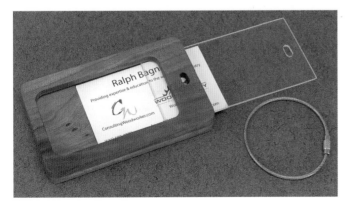

Fig. 15: Glue the wood sections together and slide the card and plastic window in from the top.

Fig. 16: Run the cable through hole in the tag to secure the parts in place. Removing the cable allows for replacing the card.

Wine Bottle Spaghetti Measure

Fig. 1: This project shows how using the same programs with different blanks can change the entire look.

Another great gift idea that has a practical side is this spaghetti measuring tool. The large holes are sized to measure dry spaghetti based on the number of people being served. There will be a little bit of carving to label the portion sizes, and the overall shape can be customized as well. This one happens to be shaped like a wine bottle, but you could choose most any shape you like. I've made this project before using a single piece of wood, but by combining the capabilities of the CNC toolpaths with different material choices, you can create unique designs without changing the programs (Fig. 1).

Supplies

◇ 14" x 4" x ½" (360 x 100 x 12mm) various glued-up blanks (see page 122 for details)

◇ 14" x 4" x ½" (360 x 100 x 12mm) contrasting veneer pieces (see page 122 for details)

◇ 60° V-bit

◇ ¼" (6mm) straight bit

1. The downloadable .dxf file available for this project (see *www.foxchapelpublishing. com/foxchapel/cnc-machining*) has the bottle shape and the measuring holes already laid out. It can be imported into a 14" x 4" x ½" (360 x 100 x 12mm) job that you set up, with the vectors joined and then centered on the panel. The Roman numerals are not included for two reasons: first, you may want to choose your own font style, and second, as mentioned before, text imported as part of a .dxf file is treated within the program as simple vectors. Text created within the CAD/CAM program is seen and treated

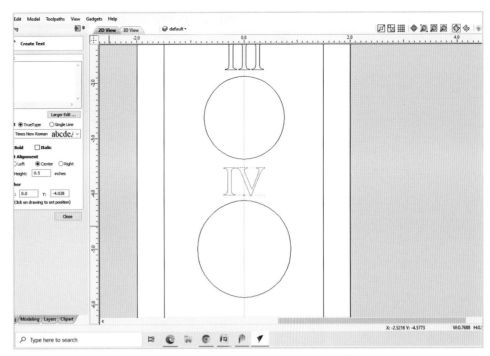

Fig. 2: Use the text functions within the CAD/CAM software to add text like these Roman numerals to your projects. This will make them much easier to move, scale and rotate as needed.

as text, so it becomes much easier to work with. The text in Fig. 2 is intended simply to identify how many servings are measured by each hole. I have chosen to use Roman numerals to enhance the Italian "feel" of the project. You can use a different font or even numbers to suit your design. Just remember to select TrueType for the font so it will carve properly.

2. The spacing between the holes is ¾" (20mm), so the numerals can't really be more than ½" (13mm) tall. You will be easing the edges of the holes after cutting, and you do not want the roundover to cut into your numerals. If you are making a measure out of a single piece of wood, the lettering can be fairly small. In this case, I want it as large as possible to take advantage of the material options we will be exploring. Choose a pocket tool path to cut the holes rather than a profile. This will add to the machine run time, but the pocketing will cut up all of the stock inside the circles. Alternately, if you were to use tabs to hold the waste, they'd need to be cut through later. And leaving loose plugs inside the holes would risk them being thrown.

Note in the drawing that at the neck of the bottle, the outer corners are rounded but the inner corners

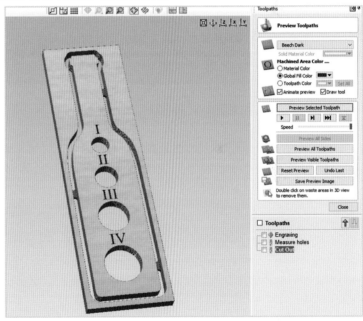

Fig. 3: The toolpath preview shows exactly how the project should look after cutting.

are not. The CNC can cut the outer corners perfectly square, but the inner corners will always have the radius of the bit, so I drew the insides square and rounded the outsides to ⅛" (3mm) radius. This is just a small time-saver for the drawing part but not a representation of how the final cut will look. You can see the results quite clearly in the simulation of the project where all of these corners will have a small radius (Fig. 3).

Fig. 4: Carving through a thin veneer exposes the contrasting wood underneath to highlight the lettering.

Making the Blanks

3. The milling steps for this spaghetti measure are even less complicated than for the sign or luggage tag projects. It is a simple project from a machining standpoint. We can take it to the next level with some creative stock choices. One of the reasons for sizing the numerals as large as possible is to create a two-tone effect using veneer over a contrasting wood (Fig. 4). If you make the core of your measure in walnut and veneer it in a holly or maple, the Roman numerals you carve into it will be highlighted as if you had painted them. You can also go the other way and use a dark veneer over a lighter core. Using very thin veneers, making the numerals as large as possible, and using the 60° V-bit will make these details stand out very well.

4. Optional: Another idea would be to glue three pieces to form the blank with a dark center and light ends or a light center and dark ends. You could position the separation lines between the sections to make the contrasting center look like the label on the bottle. You can edge-glue three 4½" (115mm)–wide sections with the grain running side to side, but but gluing two ends together like this is the weakest possible glue joint and would fail soon after you made the measure. A finger joint bit or simple tongue-and-groove joint would strengthen this spaghetti measure for generations to come. Plane this blank down to the same ½" (13mm) thickness, and both types of blank can be milled using the same program. Choosing your stock with care can add great interest to your CNC projects and allow you to make multiple custom pieces using the same G-code programs.

Considerations for Choosing the Start Point

The X- and Y-axis coordinates that the CNC machine works from are referenced as dimensions from zero in each axis. Positive numbers are generally easier to work with, so the datum point where we begin our CAD/CAM programming–which is always zero in both X and Y (0,0)–tends to be in the front left corner, with all measurements working up and right from there. Computers, on the other hand, do not work this way. Positive or negative numbers are no different inside the processing chips. That is why you get to choose the datum point on your program at the very beginning. It can be in any corner or in the middle.

As we saw when making the clamping disks (see pages 57-61), I find it easier to work from the center point when I need to run a program in a specific spot (Fig. A). This may be some section of an oddly shaped leftover or aligned to specific points on the blank. You may want to highlight an interesting grain pattern or, as with this project, line up the contrasting materials so that they end up in the correct location on the finished part. The 0,0 position on our wine bottle shape is basically at the top of the second hole. Knowing this allows for centering the bit accurately to have the "label" end up where needed. In addition, if your blank is not exactly the size you had planned on, working from the center makes it easier to keep the program centered on the actual stock.

With my other blank, the veneer did not quite cover the entire blank. To compensate, we may need to shift the start point slightly to ensure that the cut stays within the veneered area. You can do this accurately with a cardboard template, which you can easily make with your CNC. You can output just the cutout toolpath or simply run the actual program with cardboard in place. You obviously do not need to run the engraving program, but you will need to set the height of the bit to a ½" (13mm) gauge block to prevent cutting into your spoil board. Once you have the template, you can use it to find the best position on your blank and mark the start point off of that.

Throw away the inside part and the "waste" cardboard from the outside becomes a stencil that you can move over the stock until you find the best look. This technique can be especially useful if there is a knot or grain feature in your blank that you want to highlight (Fig. B). I have used this method successfully many times with highly figured wood to find the best possible section to be milling. Getting it right the first time saves a lot of money when working with expensive figured woods.

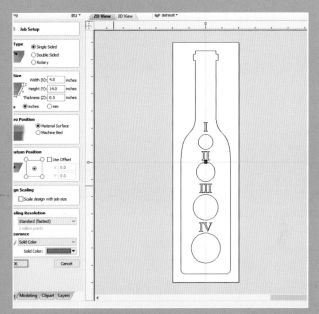

Fig. A: The starting point of any program should be selected to make the machine setup as easy as possible. Here, it's roughly centered at the top of the second pasta hole.

Fig. B: A negative space window template makes it easy to select the best possible location on the stock to set the start point.

Fig. 5: Run the text carving G-code first, then change bits, reset the Z height and run the cut out program using the same X,Y start point.

Milling the Spaghetti Measure

5. Start with the engraving program to cut the Roman numerals into your blank. Set the Z height right to the top of the stock so that the carving will be the exact depth you called out in the program. The depth of the carving is the critical thing here, not the thickness of the part. The letters are pretty small. When the program finishes, carefully inspect the Roman numerals. If they seem too shallow, you can simply reset the Z-height a few thousandths lower and just restart the program. There have been times that doing this has saved a project for me. You will never be able to perfectly realign and re-cut the lettering once the part is removed from the machine,

so be sure it is what you want before moving on to the next steps. Always take caution when doing this: if you go too deep, it cannot be corrected. Only drop the Z-height by a few thousandths at a time.

6. The cutout program is next. The X,Y start point stays the same; we just need to reset Z for the ¼" (6mm) straight bit once it is mounted (Fig. 5). This is a case where setting the Z-height off setup blocks is useful.

The blanks I glued up for the measures were not exactly the same thickness. One was very nearly ½" (12mm) thick and the other nearly 9⁄16" (14mm)

Fig. 6: Exercise care to prevent cutting too deeply into the spoil board. Otherwise, you'll need to plane off this damage, which will shorten the spoil board's useful life.

because of the added veneer. If both were milled with Z set to the workpiece surface and using ½" (12mm) as the depth of cut, one would have cut perfectly, but the thicker one would not have cut all the way through. If the cut depth had been set to %6" (14mm), both would have cut through, but as the thinner part was milled, the spoil board would have been cut into by ⅟16" (1.5m) (Fig. 6). The best solution is to program the cut depth to %6" (14mm), setting the Z-axis for both blanks using setup blocks stacked to %6" (14mm).

By choosing the best way to set the Z-height for each individual G-code, we get the best results with the least damage to the spoil board (Fig. 6).

7. As the program runs, note that because we programmed the measuring holes as pocket toolpaths, the material inside the hole gets cut into chips. We could have saved some run time by simply cutting the circles as profiles and offsetting to the inside. The problem with this is that we would have had tabs to deal with, or we would have risked the plugs getting caught in the bit and possibly flung out. Unless the holes are large, using the bit to reduce the waste to chips makes cleanup easier, especially if you have your dust collection hooked up.

Custom Coaster Set

Fig. 1: This coaster and tray set uses a fixture for two-side machining and bits for multiple purposes.

In this project, you are going to make coasters that have shallow pockets with smoothly rounded bottom edges and custom graphics carved in. These coasters can be highly customized, so they can be sold at a premium or tailored to be the perfect gift. We will make a tray to hold the coasters. This tray will be milled on both sides and mitered for assembly, all while on the machine. The techniques in this project can be used to expand your work capabilities and should inspire you to become a more efficient programmer.

Supplies

◇ 12" x 4" x ¼" (300 x 100 x 6mm) solid wood (2 per set)

◇ 10" x 5" x ⅜" (250 x 125 x 10mm) solid wood (2 per set)

◇ 14" x 6" x ¾" (360 x 150 x 19mm) MDF or particleboard

◇ 4½" x 4½" x ¼" (115 x 115 x 6mm) plywood

◇ 90° V-bit

◇ ⅛" (3mm)-radius dish-carving bit (similar to Freud #19-104)

◇ ⅛" (3mm) straight bit

◇ ¼"-20 (M6 x 1mm) threaded insert (2 total)

◇ ¼"-20 x 1" (M6 x 1mm x 25mm) nylon bolt and washer (2 of each)

You'll be using a number of different programs to make this coaster set: three to make the coasters and five to create the little tray that will hold them. On the other hand, you will only have the sanding and assembly to do after the milling is completed. With this many steps, staying organized is important. Let's take them in order.

Making the Coasters

1. We'll mill the coasters themselves from ⅜" (10mm)–thick solid wood stock. They are 4" (100mm) square with radiused corners. There will be a ⅛" (3mm)–deep pocket in the middle to hold a glass and contain any drips. To make cleaning them easier, you will radius the bottom edges and then carve your own custom design in the bottom of the pocket. Once the pockets are milled, carve a design of your choosing. This can be the date of a wedding, a favorite saying, or, in this case, a company logo. This will need to be treated and programmed somewhat differently than the carving we did on the Carved Sign we made earlier in this book (see page 84). After all that is done, the coasters can be cut out (Fig. 2).

2. If this were a plywood project, I would just put all four coasters on the same drawing and do them as a set. Solid wood is a different matter. You could glue up a panel of the needed size, but it is a lot easier to simply write the programs to work with common lumber sizes. Program the two coasters to be cut from a single 10" x 5" (250 x 125mm) blank. If you buy your lumber already milled in a "one by" format, a 1x6 will come 5½" (140mm) wide, which will work fine. You just need to cut two 10" (250mm)–long sections and run the program twice. You could just run all four in a 20" (500mm)–long board, but the 10" (250mm) drawing will work on any size CNC.

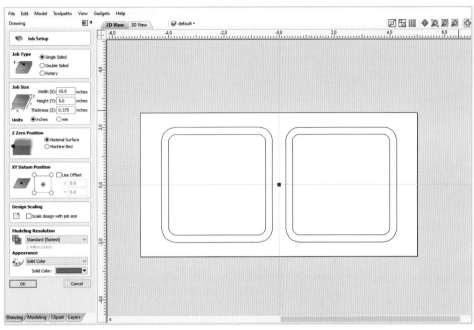

Fig. 2: The job setup for the coasters.

3. Create a workspace 10" x 5" (250 x 125mm) with a ⅜" (10mm) depth and import the "Coaster.dxf" file from *www.foxchapelpublishing.com/foxchapel/cnc-machining*. The drawing includes one pair of the coasters and two views of the tray. The coaster drawing could not be simpler; it's just a 4" (100mm) square with ½" (12mm) radiused corners for the coaster itself and a 3½" (89mm) square with ¼" (6mm) radiused corners for the pocket. Delete the tray sections labeled "Inside" and "Outside" from your CAD/CAM workspace (not on the .dxf), then highlight the vectors and be sure they are closed by using the "Join Open Vectors" command as we did when programming the arched clamps (page 63). Center the coasters on the board.

4. The carving shown in this project is not included; you get to add your own. I am using the Woodcademy social media logo here (Fig. 3). It is round and will fit nicely inside the pocket. Your image need not be round; I have done linear logos set diagonally within the pocket that turned out fine. The logo is an image file, so it has to be imported that way. In Fig. 3, the logo is obviously too large and needs to be scaled to fit and centered on one of the two coasters. You can change the size by entering the scale numerically, but it may be easier to grab one of the corner tabs and manually size it.

5. Generally, the corner tabs will reduce or enlarge objects without distorting them. Remember that you will be leaving a radius along the edge of the pocket, so keep the image at least ¼" (6mm) off the pocket edges. Most CAD/CAM programs will assist you in centering the image, snapping it to the lines or points, but even if you have to move it manually it is pretty easy to get close enough.

6. Now the image needs to be converted from rasters to vectors just as we did for the Logo Luggage Tag (see page 110). The logo is simple, and you just need to trace around the blue sections. As before, you can reduce the number of colors to two: blue and white. The software can now automatically trace around the outline of the blue parts. You can preview this and, once you are happy, just select and delete the image. The outline that remains is made up of the vectors needed for carving. You can see it in black in Fig. 4. The software uses the red shading to indicate which parts of the image have been selected to trace, which is why the blue logo image looks pink in the photo. Copy and paste the vectors into the second coaster in the workspace.

7. While you would generally program your milling steps from the center outward, you do not want to mill the logos first because they are supposed to be cut into the bottom of the coaster pocket. Select the pockets first, and program them to be cut to a depth of ⅛" (3mm).

You will be using a small dish-carving bit here, which is an unusual bit (Fig. 5). You could use a

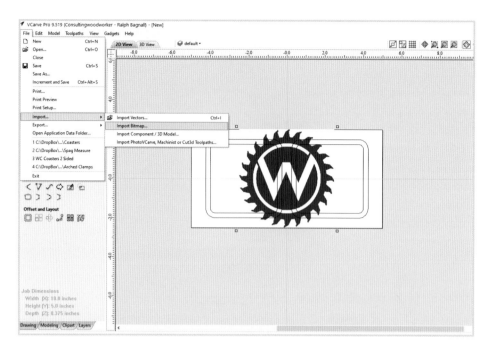

Fig. 3: Logo image imported into the coaster job space.

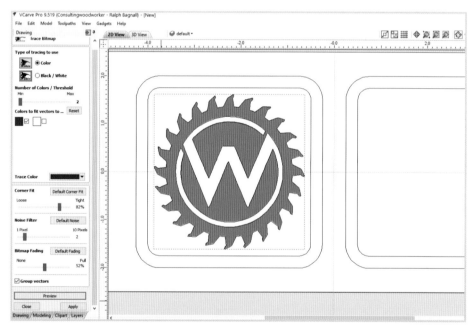

Fig. 4: The logo image has been traced and new vectors applied to the outlines.

¼" round-nosed bit, but that requires many more passes to make the bottom flat, and even then, there will be tiny ridges of wood like waves on the pocket bottom that you will need to sand out. The dish-carving bit has a flat-bottom in the center, but the carbide tips are radiused. This is so it can mill very flat pockets but the outside edges will always be radiused the same as the carbide. You will be using a bit with a ½" (12mm) overall diameter and a ⅛" (3mm) radius at the tips.

Adding Form Tools

Most CAD/CAM programs will allow for adding custom tool sizes or shapes to the database. Adding a new diameter of end mill (straight bit) is usually very simple. You can copy and modify an existing bit or specify a new tool. With these options, the parameters are known, and you only need to enter the actual dimensions. Adding a custom tool shape is a bit more involved. I will show you how it is done in VCarve, and I suspect most other software packages are similar. You should verify the process in the tutorials for your brand of CNC/CAM software.

To create a **form tool**, the computer needs a drawing of the tool to define how it will cut (Fig. A). With VCarve, you need to draw the right half of the bit's cutting profile (and only the right half). Once that is drawn and selected, open "Tool Database" and select "New Tool." This will open a menu for the type of tool to be created. If the vectors for the right half of the bit are correct, choosing the "Form Tool" option will open the dialog box with the full bit profile already drawn in Fig. A. If your vectors are not correct, an error message will appear. Your vectors should be the carbide outline of your bit from center bottom to top right, without any center line or return at the top, and they need to be selected before opening the "Tool Database."

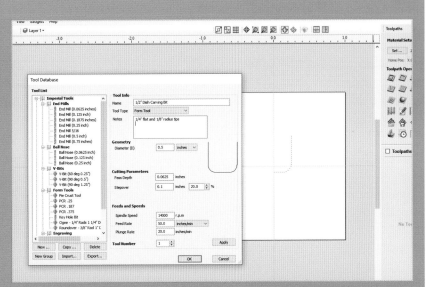

Fig. A: You'll need to create a vector drawing of the bit profile in order to define the new form tool.

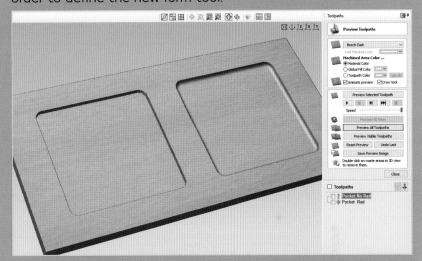

Fig. B: Add the new form tool to the tool database. The profile drawing allows the toolpath preview to accurately show the shape it will cut.

Give your custom tool a name you will remember. You can also enter notes to call out the details. This is especially important for bits that are unusual and that will be rarely used. Do not expect to remember the details a year from now.

Enter the proper dimensions for the bit you are creating. With this bit, the **step-over** needs to be specified differently than usual. Step-over is the amount of overlap between passes when the bit is pocketing. With flat-bottom bits like end mills, this is normally 40% of the bit diameter to ensure a smoothly flattened bottom.

The dish-carving bit is ½" (12mm) diameter overall, but the flat portion at the bottom is only ¼" (6mm) wide due to the radiused tips. Enter 20% to get ³⁄₃₂" (2.5m), which is correct for a ¼" (6mm) flat-bottom bit. The two pockets in Fig. B show the difference: the left uses a ½" (12mm) straight bit, and the right uses the ½" (12mm) dish-carving bit. You can see that the stepover is tighter on the right and that the inside edge of the pocket is radiused. This is exactly what you want.

In any project that may be used for food or otherwise need to be cleaned regularly, you should add a radius to the pocket bottom. The radius is much easier to get clean than a sharp corner. I have used it with bread dipping boards, snack trays, beer flights, and other projects. The dish-carving bits come in many sizes and with various radii at the tips, making it easy to choose the best setup for your project.

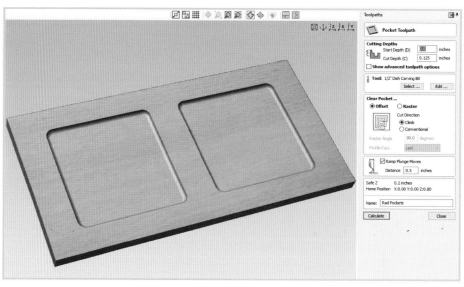

Fig. 5: The preview shows the pockets as milled using the dish-carving bit you entered into the tool database.

8. Select your carvings and begin entering the toolpath data. There are a few differences here from what you have done in previous projects that you need to carefully consider. Fig. 6 shows that 0.125" (3mm) has been entered into the "Start Depth" box, which hasn't been used up to this point. This setting is where you tell the program that you want to carve these logos at the bottom of the pocket (Fig. 6).

Also, check the "Flat Depth" box and enter 0.0312" (1/32"/0.8mm). This sets a hard limit for the bottom of the carving. Remember that the V-bit wants to cut down until it touches both sides of two lines. Some of the logo's sections are quite far apart, so without limiting the cut depth, the machine will cut right through the stock and into the spoil board. This means that there will be relatively large flat areas in the carving. It will take some time to clean them up using a V-bit, as the pointed tip of the bit takes time to flatten a larger area.

You can select a flat-bottom bit in the "Use Flat Area Clearance Tool" box, and the software will automatically create two toolpaths: one using the flat-bottom bit you choose, and a second that uses the V-bit to carve the areas not covered by the larger bit. This is often well worth doing, and you will be using

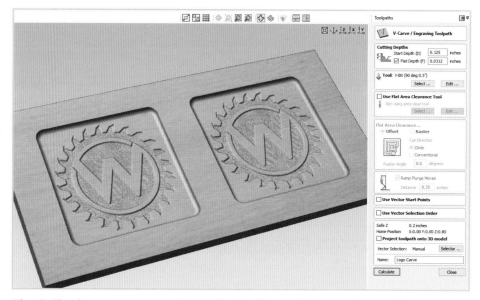

Fig. 6: The logos are carved into the bottom of the pocket. This is done by entering the depth of the pocket as the start depth at the top of the toolpath dialog box.

different tools to mill pockets in the Grapevine Relief Carving project later (see page 150). In this case, it adds another tool change, and there will be a visible difference in the bottom of the carving between where the two bits cut. Using just the V-bit will keep the texture of the carving consistent, which will mean a lot less sanding later.

9. Now you can enter the information to cut the coasters out from the blank. As before, tabs can be used to hold the parts in place, but I want to show you a technique called **onion skinning** (Figs. 7 and 8). This is where you cut parts almost all the way through, leaving about 1/32" (0.8mm) or so of material, like the skin of an onion. Since all the edges of the coasters will be visible in the final product, using the onion skin technique will reduce the amount of clean-up sanding you'd need to do if you had used tabs instead. The skin is thin enough to simply sand off. So set the cut depth to 0.344" (¹¹⁄₃₂"/8.7mm) rather than the 0.375" (⅜"/9.5mm) thickness of the stock (Fig. 7).

10. You may have noticed that the bits often leave marks where the cutout starts and ends. This is from the bit and machine minutely flexing as the cut begins or ends. It is minimized but not eliminated by ramping the beginning of the cut. I usually do not worry too much about this marking, but again, the coasters will be visible all around.

To correct the issue, use "Lead Ins" and "Lead Outs" on your machine (Fig. 8). **Leads** allow you to instruct the software to begin and end the cutout beyond the boundaries of the part, so any bit flexing does not mar the part. You can select straight or curved leads, specify how far out to start and end, and even add ramps within the lead depending on your software.

11. Don't forget that leads require extra space, and you need to be able to ensure that the lead for one part will not cut into the part next to it. You can typically control this manually, but when trying to nest parts tightly on a panel, leads can greatly reduce the number of parts you can fit, and you will have to manually check the leads for each part.

You will need to output G-codes for three different programs: the dish-carving bit, the V-bit, and the

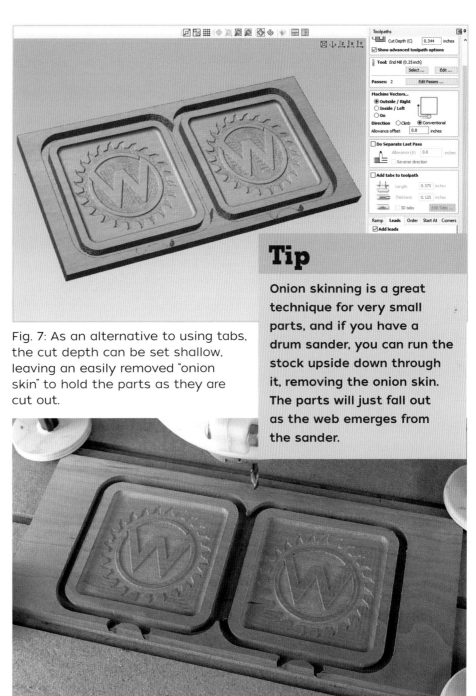

Fig. 7: As an alternative to using tabs, the cut depth can be set shallow, leaving an easily removed "onion skin" to hold the parts as they are cut out.

Tip

Onion skinning is a great technique for very small parts, and if you have a drum sander, you can run the stock upside down through it, removing the onion skin. The parts will just fall out as the web emerges from the sander.

Fig. 8: Comparing the machined parts to the toolpath preview above shows just how reliable the preview is. It is a tool that can save a lot of wasted stock. Notice also the pairs of curved leads at the bottoms of the tool-paths.

¼" (6mm) straight bit, and then run all three in order on each blank. I often name the programs with a reference to the bit being used and the order, if it matters. For example, I saved "Coaster 1 Dish," "Coaster 2 90 V," and "Coaster 3 Cutout 1-4," so that a year from now, I will know which bit to use and in what order.

Making the Tray

12. Like the coasters, the tray will be made from two pieces of stock, each containing one side and one end of the tray. In this case, however, you will make a fixture and be milling both sides of the stock (Fig. 9). This is a highly valuable technique to learn. Please bear in mind as you work through this lesson that the order in which you write the programs will not be the same as the order in which you run them on the machine, so naming the various programs will be important. Let's look at the steps in the order they will be performed before we proceed further.

Fig. 9: An indexing fixture allows for milling both faces of a part so all features are aligned accurately.

   A. The pieces of 12" x 4" x ¼" (300 x 100 x 6mm) stock will be clamped to the spoil board using your disc clamps, and they have ¼" (6mm) index holes milled into them.

   B. The 14" x 6" (350 x 150mm) MDF will be clamped to the spoil board and has ⅜" (10mm) insert holes milled into it by the CNC. This part must not be moved until all other parts are milled. It is the indexing jig.

   C. The threaded inserts are mounted to the indexing jig in the ⅜" (10mm) holes.

   D. The pieces of 12" x 4" x ¼" (300 x 100 x 6mm) stock are secured best face up to the indexing jig using nylon screws in the index holes.

   E. The custom carving is completed on the side section of both pieces of 12" x 4" (300 x 100mm) stock.

   F. The stock is flipped end-for-end on the indexing jig, the inside faces are milled, and the pieces are cut from the stock.

13. Create a job size of 14" x 6" x ¾" (350 x 150 x 19mm). This will be a fixture board for locating and indexing the stock as it gets flipped over. Import the "Coasters.dxf" file into the CAD/CAM program, this time erasing the square coasters and the part labeled "Inside." Join the vectors on the remaining drawing and center this onto the working space in the

software. You can also delete the word "Outside," as it is just there as a label. This drawing represents one side and one end of our tray. The side area is the rectangular part up to the dotted line; the end has the notch cut out of it. When you assemble the four parts, these faces will be those on the outside of the box. The two ¼" (6mm) holes at the ends should be grouped together, and the two ⅜" (10mm) holes should be made into a separate group. The outline of the tray parts will not be used at all in this case, as they are just for reference.

14. Import the custom graphics you want to carve onto the outside of the tray. It will need to be sized and moved to fit into the rectangular side of the drawing to the left of the dotted lines (Fig. 10). There is not a lot of room, so be choosy about what you try to include. Fig. 10 shows the main logo for Woodcademy being imported. Once the graphic is sized and located, you can begin to toolpath this part of the project.

15. You will need to mill these parts on one side, then flip them over and mill the other side. To do this, you need some way to ensure that the parts are in exactly the same X,Y position during all of the milling steps. By programming everything off of the same drawing and using the same zero point throughout, we can guarantee that everything will align as expected. This is largely done by the indexing jig, so start with that.

The indexing jig's threaded inserts will both locate and secure the stock when working on both faces. The drawing has both ⅜" and ¼" (10 and 6mm) circles drawn concentric to each other. The inserts I had on hand use a ⅜" (10mm) hole. You can edit the hole size if needed for whatever inserts you have at your disposal. Remember that these are going into ¾" (19mm)–thick stock even though the parts will only be ¼" (6mm) thick. Just use a straight bit in a pocket toolpath to create these holes. This is the only milling that is done to the MDF fixture.

I named this program "Tray 2 Jig 3-16" so I know what it is for and which bit to use. The ³⁄₁₆" (4mm) straight bit will be used for other things, so I programmed it here. A ⅛" or ¼" (3 or 6mm) bit would work as well, but just be sure to record which one you used.

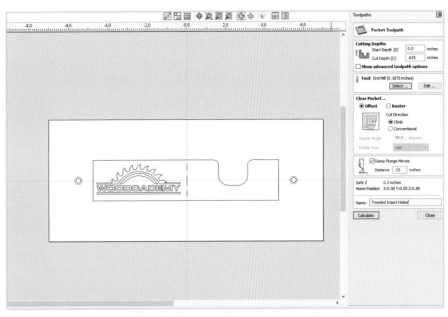

Fig. 10: The holes for the fixture and the part are from the same drawing so we know they will align as needed.

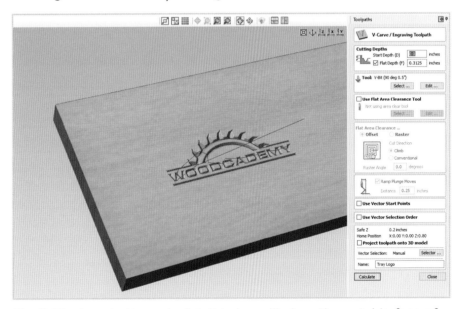

Fig. 11: The logo is the only detail being milled on the outside face of the trays, but it still needs to be accurate.

16. Select the ¼" (6.35mm) holes in the drawing and click the "Pocketing Tool Path" button. Set your depth of cut for ¼" (6.35mm) to match the stock, and select a ³⁄₁₆" or ⅛" (5mm or 3mm) bit to mill the holes. Use a smaller bit so you are not using a ¼" (6mm) router bit as if it were a drill bit. This is a standalone program. It will actually be run first to make the index holes in the stock before the indexing jig is mounted. I saved this one as "Tray 1 Index 3-16."

17. Program the carving with the V-bit. This too is a standalone, single-task program. This program is set up to use a 90° V-bit and will be milling just the logo (Fig. 11). The outlines of the parts are only really there for reference at this point. They help ensure that the holes and logo are in the correct positions. The notch in the end parts is open at the top, so that verifies that our logo is not upside down. Program the logo to ¹⁄₃₂" (0.8mm) deep as before, and while this logo image is small enough that we do not actually need to set a flat depth, you may want to set this depending on the image you're actually carving.

18. So far, we have made a fixture and decorated the parts. Most of the actual work of making the tray is done on the inside faces, and that is where the final part of this project takes us. Set up another workspace at the same measurements of 14" x 6" (350 x 150mm), but this time, when the coaster .dxf file is imported, delete everything that is not labeled "Inside." Note that the outline is flipped from the "Outside" drawing, but the notch is still facing up. There is a long, narrow box drawn in, along with three open vectors (lines), one on each end and one in the center.

19. The open vector lines represent the center of where miters need to be cut to assemble the box. If you cut along these lines with your 90° bit at a ¼" (6mm) depth, it will actually form the miter joints for you. These cuts are made using the profile toolpath rather than as an engraving toolpath. The vector offsets are set to cut on the actual line, and although the bit can make the cut in a single pass, it shouldn't for this project. In my case, between my machine and the ¼" (6mm) shank of my bit, there is enough flex in the frame of my machine to spoil the clean, straight cut needed for a miter joint. You can eliminate this flexing by simply reducing the pass depth of the bit, but that will create two equal passes. It's best to cut most of the waste in one pass and make a light cleanup pass to neaten

the cut without flexing. This can be done by editing the passes. Most CAD/CAM projects allow for this in some manner. Fig. 12 shows a dialog box that allows you to choose the total number of cuts and even set the final cut depth. Select to make a final cut at ¹⁄₁₆" (1.5mm) deep. Once this step is completed, your parts will be accurately mitered and ready for assembly right off the machine.

20. The box drawn along the bottom edge of the sides in the "Inside" drawing creates a rabbet to trap the bottom of the tray and is drawn to ¼" (6mm) wide.

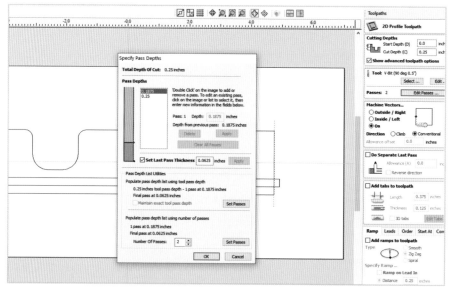

Fig. 12: The typical pass depths are modified to create a shallow cleanup cut at the end of each miter cut. This makes the miters much more accurate because it prevents unwanted flexing of the bit or machine.

If you measure the actual thickness of the stock you will be using for the base, you can adjust the width in the drawing to make for a better fit. Mine was much closer to ³⁄₁₆" (5mm) thick, so adjusting the rabbet width eliminated unsightly gaps where the bottom sits. Because of this, a ¼" (6mm)–diameter bit cannot be used to cut the rabbet. It may require a ³⁄₁₆" (5mm) or even a ⅛" (3mm)–diameter straight bit. Either is fine, and it can be left in place for the final cutting of the part (Fig. 13). The stock is only ¼" (6mm) thick, so cutting the parts out with a smaller bit is not an issue.

By selecting the ⅛" (3mm) bit, if you ever cut this job again and need to reset the rabbet width, you can do it more easily. Simply select the box and either scale it or, in this case, adjust only the

Y-dimension. In many programs, small changes to drawing vectors, like adjusting the width of this rabbet, can be updated automatically by simply clicking on the "Recalculate Toolpath" icon. If the bit is too large for the new rabbet size, the recalculation will fail. By choosing the ⅛" (3mm) bit from the beginning, you retain maximum flexibility. Again, bit choice matters.

21. The cutout toolpath is no different than any other you have previously done. The only difference here is that you will be using tabs and placing them along the bottom edge of the parts, because that edge will be hidden after assembling the tray. Also notice that two tabs are included on each part. This is because the V-bit should cut all the way through when it makes the miters, so even though the ⅛" (3mm) bit will not be cutting between the two parts, they will be separated and need more tabs on top and bottom to secure them independently (Fig. 13). I saved the G-code files as "Tray 4 Miter 90 V" and "Tray 5 Cutout 1-8."

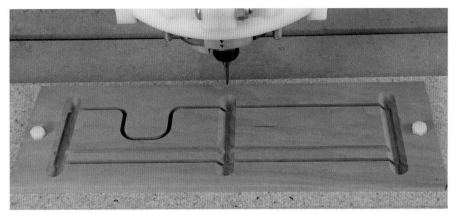

Fig. 13: This photo shows half of the tray parts that are mitered, slotted, and cut out. They're ready for assembly. The logo carving is under the right side of these parts.

Assemble the Tray

22. The entire point of this project is to show how being thoughtful in designing and programming projects can minimize the work needed after machining. In the industry, these are called **secondary processes**. In this case, there is very little to do off the machine (Fig. 14). A piece of ¼" (6mm) plywood needs to be cut to 4½" (114mm) square. The coasters and tray parts need to be separated from the web and sanded.

Tip

Remember to save the G-code files with appropriate names, especially if using many programs and bits. This project uses eight separate G-code programs and five different bits. A year from now, you won't remember the order of all the operations. Logical naming conventions keep you from having to open all the programs and check the tooling or making mistakes by being confused about which order to run them.

Fig. 14: The tray can be assembled pretty much right off the machine. The miter angles are as accurate as the bit, and the tray will only need to be lightly held together as the glue cures.

23. The coasters need no assembly, and the four sides of the tray just need to be glued together with the base trapped inside them. You can use tape to clamp everything together. Once the glue is dry, finish the parts as you wish. As coasters, they are there to catch drips and spills, so they should be protected with several coats of a waterproof finish such as polyurethane.

CHAPTER 6

INLAYS AND CARVINGS

So far, we've looked at some practical applications for CNC machines, from making their own parts to manufacturing products. In this chapter, we will be exploring the more decorative capabilities of the CNC machine. From inlays to carvings, the CNC gives you the precision and control to create amazing pieces of art.

Box Lid Fox Inlay

Fig. 1: Making the sliding lid on this box will show how easy inlays can be made on the CNC, while also using some bits in unusual ways.

Inlaying is the process of decorating a surface by cutting a shallow pocket into it, and then cutting an exactly matching veneer plug to completely fill the void. If done well, there are no gaps anywhere around the inlay. Inlays have been done for thousands of years by master artists. I have a very hard time completing all but the simplest shapes on my own. The CNC, however, is fully capable of creating matched shapes with a high degree of accuracy. Once you know the tricks, you can take your woodworking projects to the next level.

This particular project is a lid for a box that has already been built. The box is made from maple and the lid from cherry, so it certainly would look nice without decoration. Adding an inlay, however, will make this nice simple box into a work of art (Fig. 1). The inlay could be an image of most anything. A card box can be decorated with inlays of the four suits, or a pencil box with a student's initials . . . whatever works for your needs. There are some limits to what is possible and some tricks to get the best results, which I'll discuss as you work through the steps.

Supplies

◇ 9" x 4 ½" x ¼" (22.86 x 11.43 x 0.64cm) solid wood panel

◇ 15° engraving bit (similar to Freud #70-103)

◇ ⅛" (0.32cm) radius dish-carving bit (similar to Freud #19-104)

◇ ¼" (0.64cm) straight bit

Creating the Inlay Image

1. This particular box needed a lid 7¾" by 3 ¹¹⁄₁₆", so I first drew a rectangle of that size. The lid needs a finger pull at one end to make it easy to slide. I drew a 1" (2.54cm)–long line ¾" (2cm) in from and parallel to the edge that a bit will follow to form a shallow notch. There will also be a small rabbet along the edges of the lid to fit it into grooves in the box, but this can be done without any additional vectors.

2. You need an image to inlay, but not every image is going to work properly. There are two major things to look for in images to be used for inlays. First, be sure there are no inside or outside corners sharper than the radius of the bit you will be using. Both the pocket and inlay will be milled from the same vectors, so their sizes will match (Fig. 2). Any inside corners that are square simply will not get cut into the pocket, so either these will need to be squared by hand or the matching corners on the inlay will need to be rounded to fit. We'll adjust the vectors to slightly smooth the square corners and eliminate the extra steps.

Second, no areas within the vectors are too small for your bit to fit into. In Fig. 3, a comparison of the pocket toolpath using a ³⁄₁₆" (0.5cm) straight bit and the smaller ¹⁄₁₆" (0.16cm) bit shows a significant difference. The ³⁄₁₆" (0.5cm) bit can get into the thin areas like the legs, but the tip of the ear will be left out. The software simply will not try to pocket into areas that the bit will not fit.

3. Whether forming inlays by hand or using the CNC, it is important to understand that the pocket and the inlay cannot be the same size; you must provide

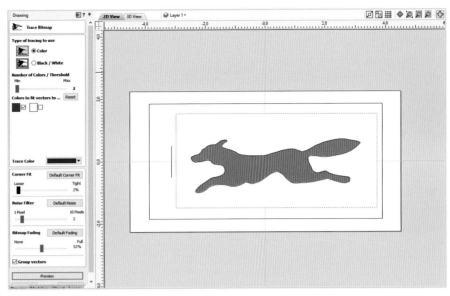

Fig. 2: The CAD drawing of the lid with the image already positioned and the vectors traced.

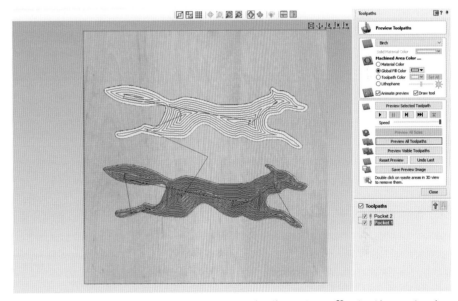

Fig. 3: These two previews show how bit diameter affects the actual shape of irregular pockets.

some clearance around the inlay or it won't fit into the pocket. Within the software, this is pretty easy to accomplish. I have made many inlays with the CNC and have found that a clearance gap of about 0.014" (0.04cm) is an excellent starting point. This could mean making the pocket 0.014" (0.04cm) larger, or the inlay 0.014" (0.04cm) smaller, but we really want to split the difference and change both by 0.007" (0.02cm). This minimizes any distortions to the shape and keeps complex inlays in their proper scale.

4. Accurately scaling a complex image like the fox within the drawing side of the software would not be easy. It is very difficult to make accurate dimensional changes by scaling an irregular figure. but fortunately there is a toolpath shortcut called an **allowance**. You'll find this setting included within both profile and pocket tool paths dialog boxes (Fig. 4). These allowances "allow" for automatically making the toolpaths around a vector larger or smaller.

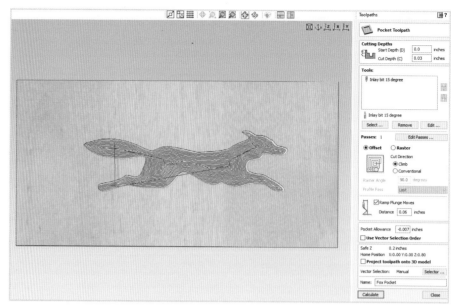

Fig. 4: The settings for the inlay pocket. Note the pocket allowance. This data can be changed to expand or reduce the overall pocket size.

Remember that the program is already shifting the actual path of the bit away from the vectors so that the edge of the bit cuts on the lines instead of on top of them. Using the allowance command lets us shift the tool paths either slightly more or slightly less as needed. This is useful for inlays, cutting hinge mortises, grommet holes, and similar areas where two parts need to fit together. Especially when you are mortising for molded or cast parts like hinges or plastic fittings, they may vary from batch to batch, and being able to simply adjust the offset instead of re-writing the toolpath is a big help. Make the pocket 0.007" (0.2mm) larger in your software by setting a negative number for the offset. A positive number moves the bit path inward, at least at the pocketing toolpath.

5. Straight bits are usually what people choose for milling the pocket and cutting out the inlay. I have found that using a bit with a small angle to it makes for a much better fit. When the pocket and inlay both have straight edges, any variations around the perimeter will be quite visible, especially when the inlay has a high contrast. Adding even a few degrees of bevel to the joint line means that the eye cannot simply see down any gap (Fig. 5). This angle also means that the inlay can be wedged into the pocket like a cork in a bottle, compressing slightly along the mating edges, which will also hide small errors for a better fit.

Fig. 5: Cutting both the pocket and inlay with even a small angle (right) hides minor gaps in the joint line beside the edges of the inlay.

6. A 15° engraving bit with a small flat tip works very well for this task (Fig. 6). The one shown here has a tip a bit less than ¹⁄₁₆" (0.16cm) in diameter so it will cut into the smallest practical corners. The tip is flat so the pocket will be smooth along the bottom. The walls of the pocket will be milled at a 7.5° (19.05cm) angle, providing the **scarf joint** effect we want where the parts meet. You will likely need to set this tool up in your database (Fig. 7).

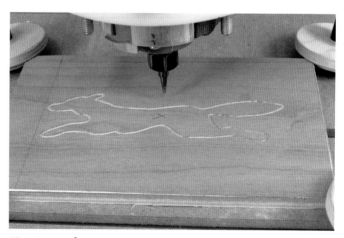

Fig. 6: A 15° engraving bit with a flat tip is an excellent choice for inlaying a shape with fine details.

7. It is categorized as an engraving bit, which allows for entering the angle like a V-bit, but also the tip diameter. This is important because it lets the CAD/CAM software know to set the toolpath as if it were a ¹⁄₁₆" (0.14cm) straight bit, instead of using the overall diameter, but the software should still show the angle in the toolpath preview if the cut is deep enough. Our pocket is only 0.03" (0.08cm) deep so the angle is not very apparent.

8. When setting up the pocket toolpath, you will have the choice of using a larger tool to remove most of the waste. This is a choice you need to make whenever the bit being used is a small diameter. Using the larger tool will shorten the runtime of the machine, but it does require a tool change. This fox image is small enough that I was able to avoid the tool change, but the pocket would not need to be too much bigger to justify the tool change due to the small bit.

9. With the inlay pocket programmed, the other details in the box can be set up. We'll be forming a finger pull and rabbeting the edge of the lid. You can save another tool change by choosing the right bit for these tasks. We'll use the same ½" (1.27cm) dish-carving bit

Fig. 7: Setting up the engraving bit in the tool database

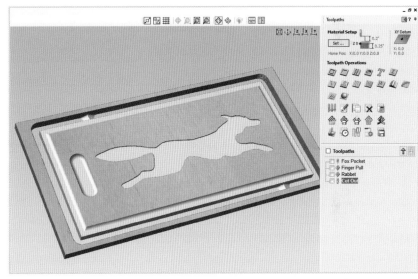

Fig. 8: The dish-carving bit can be used to create the finger pull and the radiused rabbet around the edge of the lid.

as used to pocket the coasters, but in a different way (Fig. 8). The finger pull could have been drawn as a closed oval vector and programmed as a pocket, but using a single line is fast and easy. It also allows us to change the shape of the finger pull instantly by just changing bits. A ¼" (6mm) round nose bit will make a different shape than a ½" round nose will. And the dish-carving bit we are using makes a third shape. With the single open vector, you rely on the bit diameter and profile to form the pull. It is a simple profile tool path with the bit cutting on the vector. The dish-carving bit we are using will cut a wide, flat bottom so it will not plunge well. Be sure to set the ramp length longer than usual.

Rabbeted Edges

Before plywood, drawer and box bottoms and lids were made with rabbeted edges. This allowed the panel to be thick enough to resist cracking or warping but also kept the groove in the thin box side as narrow as possible. This is the concept behind raised panels for cabinet doors.

I set it to a full 1" (2.54cm) long. The bit is ½" (1.27cm) and the line being cut is only 1" (2.54cm) long, so extending the ramp distance will actually make the bit cut the pull in one smooth pass down to the bottom. This is a great technique to keep in mind for any short slot cuts that you may need to make.

10. We need to thin the edges of the panel to fit within the rabbet. This is called "raising" the panel since it leaves the center at full thickness. Raise the lid of the box with a small rabbet around the edge. This could be done with a straight bit, but it is more pleasing

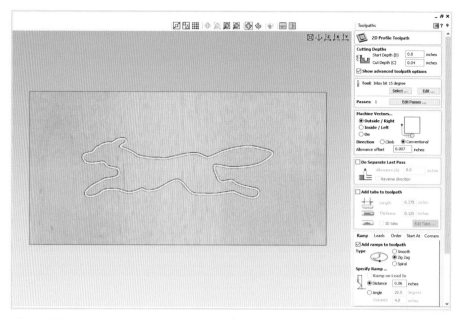

Fig. 9: The inlay is cut upside down (mirror image) so the angled edges will align with the pocket.

to the eye if the edges are beveled or rounded. Since the dish-carving bit is already in the machine, we can use it to cut the rabbet. In previous chapters, you created pockets and rabbets by drawing enclosed shapes that get milled out, but since the rabbet does not need to be wide here, you can simply make it the same as the bit radius. This means you only need to toolpath the dish-carving bit to follow the outline of the lid. This will make a ¼" (6.35mm) wide rabbet cut along the edge of the panel. The inside edge of the rabbet will have a ⅛" radius cove providing a graceful edge. The rabbet will actually be ½" (12.7mm) wide, but the outer half will be removed as the lid is cut out.

One good reason to program this way is that by simply changing the bit used in the program, you can change the size of the rabbet. A ¾" (2cm) or larger dish-carving bit will make a wider edge of the same basic shape without needing to redraw any vectors.

11. With the lid programming completed, cut out the inlay plug that will be fitted in. This must end up the exact shape and size needed, so you will use the same vectors that you used to cut the pocket. You imported the fox image for the lid, and then traced the edges to get the vectors, but you also scaled it to fit. The easy way to ensure that everything will work is to resave the lid CAD/CAM file as a

different name and work the inlay from there. Any resize or adjustment you have done to the fox image will remain but there is no danger of accidentally copying over the lid program.

12. Delete all the toolpaths and the vectors for the finger pull and lid edges. You can even reset the job size, but the image's shape and size must not be altered in any way. Cut out the inlay and use the 15° engraving bit to angle the edges of the plug. They need to be angled to fit the pocket, so our fox inlay needs to be cut as a mirror image of the pocket (Fig. 9). Keep in mind that if your inlay shape is symmetrical, flipping the image does not matter, but in either case the bottom of the veneer being cut will become the visible face.

13. Fortunately, your CAD/CAM program can flip the vectors with just a click or two. Set up the toolpath needed. It is a standard profile toolpath, set to cut just through the veneer, with the tool set to cut outside of the vectors. This part will not be held with tabs since the veneer is too thin, and removing the tabs can spoil the fit of the inlay. The veneer will need to be pasted down to a substrate for cutting, then removed to glue into the pocket (see "Holding Delicate Parts" next page).

Just as we used offset to make the pocket slightly larger, we will use them here to make the plug smaller. As before, the offset should be about -0.007" (-0.018cm). It seems wrong that the pocket and the plug should both need to be offset by a negative setting, but that is how my program works. You will want to test how your program works with a simple inlay in some scraps. This is one case where the toolpath preview function will not be able to show you the difference because it is too small to see in the preview graphic.

14. In this project, I have shown the manual method for offsetting the pocket and inlay. This is the only way it could be done when I learned it. But today, most CAD/CAM software packages have a specific toolpath option for making inlays. This function usually works by combining the pocket and cut out tool strategies and adding in the offsets for you. You still enter the amount of the offset, but you do not have to figure out if your number should be positive or negative; by choosing the inlay tool path, the program knows what you need to do.

Holding Delicate Parts

The CNC machine excels at cutting inlays, but holding the veneer for cutting on the machine is problematic. It is simply too flexible for clamping down in the usual manner and too fragile to double-face tape it to a substrate. It will cut just fine, but removing it from the substrate in one piece is nearly impossible. An old woodturner's trick is just the ticket to solve this problem.

If you have ever seen a pair of perfectly matched half columns on a clock pediment, they were likely glued together from two pieces of wood, turned as a single unit, and then separated again. It would be dangerous if these pieces were to come apart while spinning on a lathe, so the joint needs to be secure. However, it also needs to separate cleanly and without damage to the parts once it's turned.

A thin layer of cardboard glued to the fragile inlay can be split after milling to release the veneer without damage. The cardboard is scraped off after the plug is glued down.

A layer of thin cardboard is the key. The two pieces to be turned are glued together with cardboard sandwiched in between. Think of the kind that a cereal box is made from or the cardboard that a dress shirt is packaged with. The cardboard is strong enough to hold the parts when working, but it will split along its thickness when pried apart. Careful work with a chisel and flexible putty knife will release the inlay in one piece.

By gluing a thin layer of cardboard between the veneer and substrate the cardboard can be split after milling to free the veneer. The face of the veneer will have glue and/or cardboard stuck to it, but you milled it upside down, so once the inlay is glued into the pocket. All of that waste material can be removed as the veneer is scraped flat with the surrounding surfaces. I have found that spray adhesive is excellent for gluing the veneer in this way. It is not a permanent adhesive, but it only needs to hold long enough to cut out the inlay plug. It is available at most home centers, craft, and hardware stores, and is easy to use. Wood glue will work and will be scraped off anyway, but it can soak into the veneer, which may cause trouble later when applying finish to the project. Double-faced tape can also be used with the cardboard in between. Whatever adhesive method you use, just remember that anything stuck to the veneer face is best removed after the inlay is securely fixed in its pocket.

Ibis Photo Carving

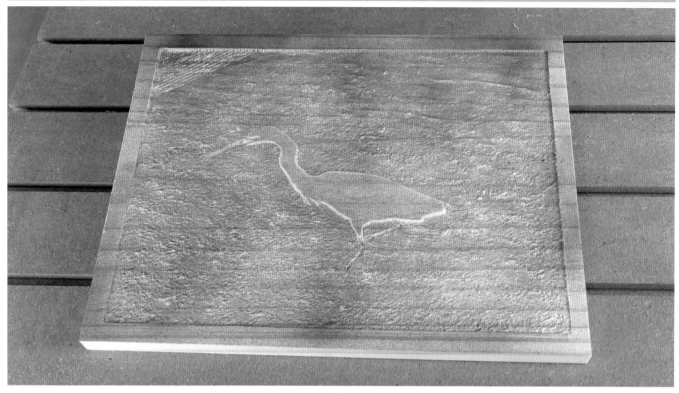

Fig. 1: Converting photographs into carvings on the CNC is a challenging but rewarding task that is very popular among users. Staining, dry-brushing, or even wood burning can help bring out details.

Taking personal photos and carving them into projects is certainly one of the top reasons that new CNC owners give for buying their machine. Naturally, this ability is marketed heavily by both machine and software manufacturers, but it is not nearly as simple to get great results as they make it seem. You certainly can get very cool results, but careful choices need to be made at every step. From choosing the original photo to selecting the best bit, through programming options and even deciding what material to carve into, each choice can greatly alter the outcome (Fig. 1).

Not all CAD/CAM software systems offer a photo carving option as we will be doing here. Vectric's VCarve Pro, the software I am using, only added it with their version 10 release in 2020. They have a standalone program called Photo VCarve to do this work that has been around for some time. There are others available, but it is not as common a type of software as a typical CAD/CAM package is because it's not multifaceted. The photo carving software is pretty limited to just the photo carving process.

Supplies

◇ 12" x 9" x ¾" (30.5 x 22.86 x 2cm) wood panel

◇ ⅟₃₂" (0.08cm) tapered ball-nose bit (right)

◇ 15° or 30° engraving bit (optional)

All of the variables are controlled by how the photo is read by the software. Essentially, it is divided into lines of varying widths. In darker areas the line is rendered wider and lighter areas are rendered as narrow lines. These lines are then basically programmed in the same way as engraving tool paths are, with the V-bit filling the width of the lines, cutting deeper as they get wider, and rising up as they narrow.

1. Choose a photo that features good contrast, strong shapes in the foreground, and clean, uncluttered backgrounds. These are the types that work better for engraving. You don't have to convert the photo to gray scale to make this work. I selected a photo of an ibis I took at the beach (Fig. 2). The subject takes up a good amount of the photo, and while our eyes see the complexity of the sand, stone, and water that make up the background, it will become a generic texture behind the bird in the carving.

2. Set the job size to the portion of the photo you wish to include. Very small carvings of this type will not work well because the lines that will be created to make up the photo simply will not have enough width to allow the bit to cut into the stock.

Since most photos are rectangular, that is how they will import into the CAD/CAM software. If you want a round or oval carving, you will need to use photo software to shape the image before importing. For this photo, I have set up a job 12" x 9" x ¾" (30.48 x 22.86 x 2cm). Import the photo as a bitmap just as you did with the logos in previous projects. This time, however, don't use the "Trace Bitmap" commands to convert the image to vectors. The photo carving software will do that for you, automatically converting the image as it tool paths.

The imported image can be scaled manually to fit the stock, or you can use the scaling commands within the CAD/CAM software to fit it to the board being carved. Nothing else is needed from the CAD part of the program, so you can move right into setting up for the toolpath.

Fig. 2: Photo choice is critical for success with this project. High contrast subjects and uncluttered backgrounds work best.

Fig. 3: Each setting will change the final carving. Experimenting with them in the toolpath review will help.

3. There are only a few actual variables to set up for photo carving, but each one can make very visible changes to the final carving, and often they need to be adjusted in combinations (Fig. 3). The program is going to carve the image in a series of straight lines with the bit moving up and down in the Z-axis as it cuts each line, very much like the way an inkjet printer works. (See the "Program Variables" sidebar on the next page for more information.)

Program Variables

Here are the variables in the VCarve photo carving routine with notes on how they control the image. Not all CAD/CAM programs will have all of these options, but if they can do photo carving, they will have similar options to work with.

1. **Start Depth** - This is zero unless the carving is to be done at the bottom of a pocket.

2. **Max Carving Depth** - This is how deep the darkest parts of the carving are to be cut. The deeper this is set, the wider the program has to separate the individual lines, so this number is always a compromise. Deeper cuts mean more detail in Z, but it also means fewer lines overall since each will be wider as the V-bit cuts deeper and resolution is lost.

3. **Stepover Retract** - This is how high the bit will lift up to move to the next line. This entry does not alter the look of the carving in any way, but the bit will lift up, move over, and drop down between each line. You want this set so the bit does not cut between lines, but the less the bit travels up and then back down, the less time is consumed.

4. **Tool** - Which bit will you be using? How tool choice alters the carving is very closely related to the max carving depth. Photo carving uses V-bits just like engraving does. A 60° V-bit makes narrower lines than a 90° bit when cutting to the same depth. At the same max carving depth, a 90° V-bit will allow for fewer lines than a 60° bit, so the 60° bit will cut an image with more detail.

5. **Raster** - This instructs the software to divide the image into lines. The angle of the lines can be controlled with the line angle slider.

6. **Hatch** - Selecting this option creates a second raster set of lines that run perpendicular to the angle selected on the line angle slider. This adds detail to the carving by working the dark and light areas from two directions. It also doubles the runtime of the carving.

7. **Selected Vectors** - In the CAD section of VCarve, repeating patterns of vectors can be added to a job. If you have these vectors set up, selecting this button will cut the image following them rather than making straight lines at a selected angle.

8. **Line Spacing** - The initial line spacing is determined by the max carving depth and the bit selection, but it can be expanded. The most lines you can have per image is 100%; anything more than that separates the lines more. This will reduce runtime but also image quality.

9. **Line Angle** - This slider is used to set the angle that the lines follow when cutting. In the center at 0, the lines run horizontally; at 90 they run vertically. The angle can be set anywhere in between. Cutting at an angle seems to fool the eye into noticing the cut lines less than if they are straight up or sideways.

4. The default settings for my photo carve were 0.1" (³⁄₃₂"/0.24cm) for max depth, a 60° V-bit, and a raster cut strategy at 45°. I left them there, calculated the cut, and previewed the toolpath. The results were not good.

5. In Fig. 4, notice how the lines are spaced wide apart. This led to poor resolution and a lot of lost detail. Since the line spacing was set to 100%, the width was due to the depth of cut and the bit itself. To learn how the settings work, try to only change one at a time. This allows you the chance to see what each setting does. I left everything alone and only changed the max carving depth, reducing it until I found what I thought was the best setting, which was about 0.04" (¹⁄₃₂"/0.08cm). When I set it much shallower than this, the carving started fading out altogether.

6. Since the bit width also affects the resolution, I tried adding some engraving bits to the toolbox. Engraving bits are a little more difficult to find than standard bits, but they are available. For a reasonable price, 30° bits are available but 15° bits are somewhat less common. Both will carve finer than the standard 60° V-bit (Fig. 5). Just be careful: many engraving bits feature a flat tip instead of a point. If the flat is 0.005" (0.01cm), it really will not matter, but wider flats will change the look of the carving.

Fig. 4: Too much distance between cut lines reduces the resolution of the image.

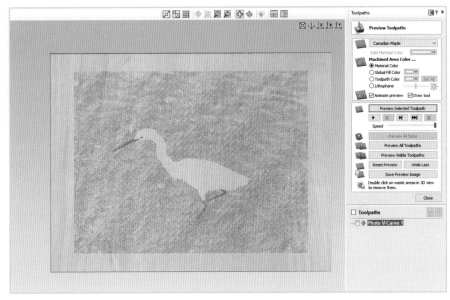

Fig. 5: Narrower V-bits allow for tighter lines and higher resolution.

7. I chose to include a tapered ball-nose bit with a ⅟₃₂" (0.08cm) ball (Fig. 6). This bit has a 6.2° angle, so it will program with much narrower lines than the 60° V-bit. It has become a much more common bit since it is well suited for milling 3D models and these bits seem to last forever.

The ⅟₃₂" (0.8mm) round tip works just as well as a pointed V-bit because this bit will be cutting relatively deep for a photo carving. Using this bit with a depth setting of 0.2" (³⁄₁₆"/0.51cm) has even better resolution than the 60° bit at 0.04" (⅟₃₂"/0.08cm).

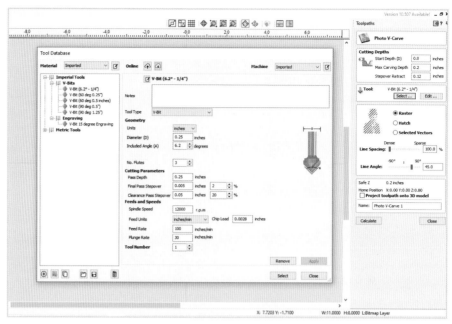

Fig. 6: Adding a new V-bit into the tool database

Settings and Tools

The lesson here is that the preview allows you to quickly try any number of different settings and tools until you get the results you want. You will come to understand the relationships between the settings and tools pretty quickly without having to cut any stock to learn from.

8. Another interesting detail I found in experimenting in the preview mode was that the "Line Angle" setting does make a difference. In the photo of the bird, the waves in the background are regularly spaced apart, as you would expect. With the line angle set at 45°, the background kept taking on a quilt pattern in the preview. Resetting the angle to 30° eliminated this, and the image looks far more natural (Fig. 7). Experimenting in the preview mode is time well spent and saves material and time at the CNC.

Fig. 7: When all the settings are correct, the image will be very clear in the toolpath preview.

9. Be prepared: carving your photo will take a lot of time. The bit will be traveling back and forth along your stock, only stepping over a small amount with each pass. You'll get more lines with a smaller bit, so it will take longer to carve. Depending on the size of the carving and the steps between passes, it can take hours. Carving the ibis image with the 60° bit took me a little less than half the time carving the same image with the 6.2° bit.

10. It's also important to consider the wood species for your carving. With hand-carving there are preferred species, and that's true here as well. Choose lighter woods with tight, closed grains like basswood, soft maple, or beech. Highly figured woods generally do not carve well, nor do open-pored species like oak. Avoid knots and other blemishes. Very dark woods may carve cleanly, but the image can be harder to see.

Grain patterns like zebrawood or tiger maple are also not good choices because the carving and the wood patterns tend to blend and distort each other.

Many plastics work well, and carving a negative image into clear acrylic can be stunning when lit from behind. This is called a **lithophane** and is very popular as a craft item or gift.

11. This is a project where using the CNC to plane the stock flat is required. With a 60° V-bit, the maximum cut is only a little more than 1/32" (0.08cm) deep. Even with the 6.2° taper bit, the cut is less than 1/4" (0.64cm) deep. Basically any variation from absolutely flat across the carving area will noticeably distort—if not completely ruin—your work with some parts too deep and others too shallow. If you are carving at the bottom of a pocket, it will already be flat so there's no need for the extra planing step.

Fig. 8: When the settings are correct, the image will be as clear as possible. Here, the bird is highly visible and even the beak and legs stand out from the waves in the background.

Safety Precaution

You don't need to watch the machine during the entire program run, but never leave the machine running unattended. Too many things can go wrong. I have even seen industrial-level machines get damaged because the operator was not nearby to shut it down in a timely fashion. As long as you can hear the machine and reach it in a few seconds to hit the E-stop, you can be working in another area of the shop. CNC machines are very loud. If a bit breaks or a part comes loose and starts moving, you will hear the change in sound instantly. Just be sure you are close enough to react. Most machines can be paused in the middle of a program if you must leave the area for a short time.

Versatility of Photo Carvings

If you have the right software, carving photos can be done as fun standalone projects or incorporated into keepsakes, boxes, or furniture parts. It does require very careful consideration of the bits to be used, the setting selected, and the stock to be carved (Fig. 8). Learning to use the preview function in your CAD/CAM software is a powerful skill that will help you learn more quickly and get better results while wasting a lot less stock.

Grapevine Relief Carving

Fig. 1: This project will mill relief carvings using profile bits to create the shapes needed.

Using V-bits for lettering and carving graphics or even photo images into your projects is fun to do and highly useful as part of your skill set. This project is going to step things up to the next level, as you will be creating actual relief carving (Fig. 1). You will do this by using standard toolpath strategies with advanced tool choices to form the patterns you want.

Supplies

◇ 8" x 7" x ¾" (20.32 x 17.8 x 2cm) wood panel

◇ ¼" (0.64cm) point-cutting round over bit (similar to Whiteside #1572)

◇ ³⁄₁₆" (0.48cm) point-cutting round over bit (similar to Whiteside #1570)

◇ ¼" (0.64cm) straight bit

◇ ¹⁄₁₆" (0.16cm) straight bit

◇ ¼" (0.64cm) ball-nose bit (similar to Freud 72-101)

◇ ⅛" (0.32cm) straight bit (optional)

1. This project starts with a 2D vector drawing. Begin by setting up a job at 8" x 7" x ¾" (203 x 178 x 20mm) and import the file "Grape Carve.dxf" into the software. The original came to me as part of an architect's drawing for some decorative panels to be carved by hand. Sadly, artistic drawings are almost never useful for creating toolpaths, even when they come as a .dxf file. I had to heavily re-work the image so the tools cut the way I needed.

2. Because this is a relief carving, most of the material in the work area will be removed as a pocket so that the details remain proud of the new background. To create a pocket with uncut "islands" inside, the islands and the added surrounding pocket borders need to all be defined as closed vectors. In this case, the leaf, vine, and grapes are the islands and the outer border surrounds all of them. This border can be rectangular as shown or any shape needed, as long as it is a closed vector surrounding all internal details.

3. This .dxf file is certainly complex enough to make grouping similar vectors a necessity (Fig. 2). First, group all of the closed vectors that define the outline of the leaf and vine. The vectors for the leaf and vine also include the small irregular circles that are inside the leaf. They need to be pocketed along with the rest. Select and group together the circles that will become the grapes. Finally, all of the lines that make up the details on the leaf and vine become a third group.

4. Select the pocket toolpath to mill down all the background areas. There are a lot of small details to be defined by the bit, so choose the smallest bit possible. I used a ⅟₁₆" bit here. You could substitute

Fig. 2: With complex drawings, grouping similar elements together is a valuable technique.

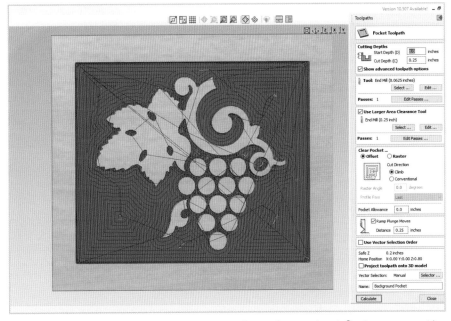

Fig. 3: When pocketing large areas, using a larger tool first can greatly reduce the milling time. Most CAD/CAM software programs can set this up automatically.

a ⅛" bit instead, but you'll lose some of the details when carving with it. Using a small-diameter bit to pocket out the large open sections of the background is a waste of time, so I suggest that you use a larger tool—a ¼" (0.64cm) straight bit—to mill out most of the waste.

Most CAD/CAM software makes this automatic. Two toolpaths are created with the larger bit removing most of the waste first, and the smaller bit only needing to cut where the larger bit does not fit between the islands (Fig. 3). Previewing the pocketing will show you that you've selected all the islands and nothing is being cut that should not be. The widely spaced blue lines show the path of the ¼" (0.64cm) bit. The rest are the ¹⁄₁₆" (0.16cm) bit cleaning out the tighter spaces.

Now you have a recognizable image but without any form or detail to it. Today, carvers often use the CNC to get to this point and then finish the work by hand. Is this cheating? It might be, but remember that before CNCs, an apprentice carver would do this sort of work and the master would only carve the fine details. I am certainly no master carver, so I used the CNC do the rest.

5. The grapes need to be rounded over to make them into hemispheres. Select the group of circles and begin a profile toolpath. The background pocket is being cut ¼" (6.35mm) deep, so set the cut depth the same. A ¼" radius, point-cutting round over bit will be your tool selection.

Most round over bits have a fairly wide flat along the tip that will not fit between the grapes. It so happens that there is a bit that can do just that. The point-cutting round over bit (Fig. 4) can be bought in several different radii. The carbide tips come to a sharp point rather than being flat, so it can cut fully around each grape circle without cutting into those nearby.

6. Point-cutting round over bits are not included in the standard tool database, so you will need to create them. This is the same procedure done when making the Custom Coaster Set (see page 126). Draw the right half of the bit and open it as a form tool. In this case, the right half is a quarter-circle with a short straight leading up from it.

Because you'll be using a ¼" (0.64cm) radius version and a ³⁄₁₆" (0.47cm) bit, you might as well draw both out and create the tools. You can draw both vector

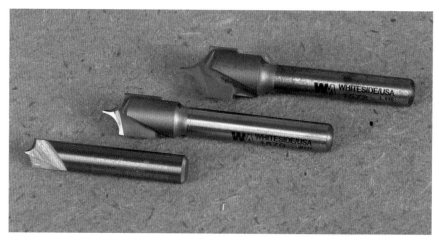

Fig. 4: Point-cutting round over bits come in several different sizes and radii.

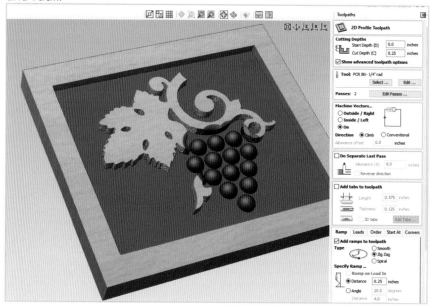

Fig. 5: The preview clearly shows how the round over bit can create perfect hemispheres.

profiles in the same job setup and then just select one at a time to create the tools. It takes a couple of minutes, but it will be very important that these are entered so that the preview image will show the carving properly.

7. Select the group of circles, set a ¼" (0.64cm) depth, and select the new ¼" (0.64cm) point-cutting round over bit. It can cut to full depth in one pass, but this sometimes leads to some tear out in the wood grain. Instead, choose two passes, one at ³⁄₁₆" (0.48cm) deep and the cleanup pass at ¹⁄₁₆" (0.16cm) deep. This should reduce the chance for tear out and clean up any that may still happen. Set the tool to cut on the vectors. The point of the bit will follow the circle and the round over will form the hemisphere inside it.

8. The preview shows the grapes formed as planned (Fig. 5). You can also clearly see that the round over will cut partially into some of the surrounding detail. This is why it is worth taking the time to create the tooling with an accurate profile. Normally, this would be a warning of a flawed plan, but you will actually be rounding over these other parts next, so there is no need to fix the grape toolpaths.

9. To round over the leaves and vines, select those vectors again, and program them using a profile toolpath. Change the depth to ³⁄₁₆" (0.48cm) deep, select the ³⁄₁₆" (0.47cm) point-cutting round over bit, keep the bit cutting on the vectors, and calculate the toolpath. The preview now shows all of the vine and leaf boarders smoothly rounded over (Fig. 6). The effect is especially effective along the vine, really defining it as an organic object. The preview also shows that the unintentional cuts from the previous toolpath are milled away so no changes are needed.

10. Now program the veining on the leaf. This could be done with a V-bit, but since we already have the ³⁄₁₆" (0.48cm) point-cutting bit in the machine, we can simply use it. Set the depth shallow since we only want to "draw" in the veins as surface detail. Again, the

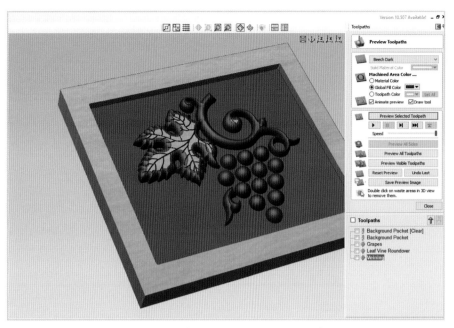

Fig. 6: Rounding the edges of the various carvings makes them look more natural, and the point of the bit can be used for adding details.

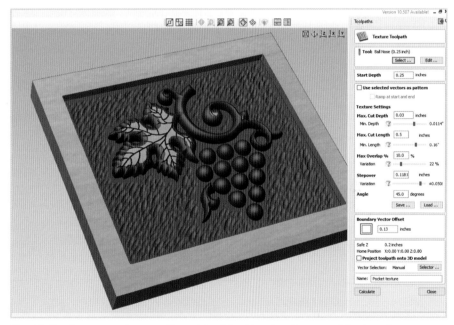

Fig. 7: Texture can be easily added to the background to lend a handmade look and eliminate tool marks.

preview feature helps here. With a V-bit, the lines get wider in proportion as the cut depth increases. The point-cutting round over bit flares out from the point into the two radii, so as you increase the cut depth, the veins become progressively wider. I found that ¹⁄₁₆" (0.15cm) was a bit too faint and ³⁄₃₂" (0.23cm) looked too deep, so I settled at 0.070" (1.7mm) deep.

11. The final step is to texture the background (Fig. 7). Many CAD/CAM programs allow for setting up

a random texture to be cut on a face. If yours does not, it shouldn't be a problem. You can just sand the background smooth and leave it as is. Adding the texture, however, covers up any variation in the Z-heights of the bits that cut the pocket, meaning the pocket does not need to be sanded as much. Only texture the bottom of the pocket. The same vectors need to be selected as when the pocket was programmed: the outer box, the outlines of the leaf and vine, and the grapes.

12. The texturing toolpath allows for creating random textures and flat surfaces, and it includes a number of variables for you to control the look. The first variable is the bit itself. You can choose any bit, and a V-bit will create a very different texture than either a straight bit or a ball-nose bit. I chose the ball-nose bit to simulate more closely the look of a hand-carved panel. The variables allow for choosing the maximum depth of the texture cuts, setting the max and minimum length of the cuts as they are randomized, how much overlap is allowed between rows, the range of step over, and the angle at which the texture is oriented.

The last option is the boundary offset. The software is always controlling the bit from the center point, so this texturing step can easily cut into the edges around and inside the pocket. Set this to be just a little more than the radius of your texturing bit, and it will prevent any accidental cutting into your detail. As always, using whatever preview function your CAD/CAM software has will get you dialed in before you even make the first cut.

13. This project uses five different bits and G-codes, and the order in which they are run matters. Make sure that you name them all so that you will always be able to tell which bits get used in what order. I named them as follows:

    1 Grape Carve 1-4 PCR
    2 Grape Carve 1-4
    3 Grape Carve 1-16
    4 Grape Carve 3-16 PCR
    5 Grape Carve 1-4 Ball

Note that the numbers 1-5 are part of the program names so there is no mistaking the sequence.

14. You may notice that the milling is not in quite the same order as you programmed the steps; the grapes are shaped first. As I was developing this project, the wood grain in the grapes kept tearing out as they were cut. To help prevent this, we added a cleanup pass to the toolpath, but cutting them first will also help. If all the wood is removed around the grape circles first, then as the round over bit crosses the grain pattern, the material is not supported. By shaping the grapes first, there is a lot more support to protect the grain. The point-cutting round over bit is more than capable of making this cut first, especially with the two passes we programmed in.

15. You then run the rest of the G-code programs in order; the ¼" (6mm) bit milling most of the pockets, the 1⁄16" (1.5mm) detailing the edges, the 3⁄16" (5mm) round over shaping the vines and veining the leaf, and the round-nose bit texturing the background. There are four tool changes to make between programs, but you should find that each of these G-codes run pretty quickly. This carving (see Fig. 8) should take about half the time to cut as the photo carving project (see page 144).

Fig. 8: The finished carving with all elements combined.

GLOSSARY

This glossary is not meant to be regarded as an official list of definitions. I have created it to clarify terms as I use them in this book. Like many sciences and professions, terms used by CNC programmers and operators have sometimes evolved to be different than the common usage. Use this glossary as a reference when reading to understand unfamiliar terms as you encounter them.

Allowance – When the toolpath is automatically offset from any raster line by a predetermined amount. Useful for inlays and for making minor adjustments to holes and pockets without needing to re-draw the part.

Bit – The cutting tool mounted into the router or spindle of the CNC machine. This may be specially manufactured for use in CNC machines but does not need to be. Almost any common router bit that is not bearing-guided can be used in a CNC. (See also tool.)

Bridge – Component that moves forward and backward along the frame of the CNC. It holds the head, which moves side-to-side across the bridge. The front-to-back and the side-to-side motions make up the X- and Y-axis, but which is which may vary between machines.

CAD – Computer Aided Design; a computer program that allows the user to create drawings in scale with dimensions.

CAD/CAM software – Computer programs that combine both drawing functions with toolpath functions to allow programming CNC machines in one software package.

CAM – Computer Aided Manufacture; computer programs that allow for associating physical milling operations to individual details contained within a vector drawing.

Cartesian coordinates – A system of locating objects within a field by defining points across a plane. First described by René Descartes in the 17th century, the system is how all machines can work accurately within a known area. Starting from a point chosen to be zero, all positions can be plotted by moving known distances in known directions. The system allows all points to be accurately returned to when needed. (See also coordinate.)

CNC – Computer Numerical Control; Guiding machine processes and movement using software programs. The term has come to mean the machines themselves even more than a method of control.

Collet – A clamping piece used to hold router bits within the spindle or router motor. Most common in the US are ¼" and ½" (6.4 and 12.7mm) diameter, but other sizes may be available.

Coordinate – The measurement of distance in any axis from the center point. Coordinates are generally used as "absolute" numbers that reference back to the zero point, where a "relative" move is covering a specified distance from the current point. (See also Cartesian coordinates.)

Datum – The anchor point of a CAD drawing or CNC program. It is defined as 0 in all three axes and all items in a drawing or movements in a program relate to it. The datum can be reset, and all other points will move in relation to the new datum. (See also start point.)

DXF file – Just as G-codes are a universal standard language to operate different types and brands of CNC machines, the ".dxf" file format allows designs from different CAD programs to be easily and accurately shared.

End mill – A rotary cutting tool used in CNC machines. The term "end mill" is usually associated with metal-working tools rather than wood, but most CNC router bits can be considered end mills.

Feed rate – The speed at which the router bit is moving across the part as it is cutting. This is usually stated in "inches per minute" or "mm per second." It is different than the RPM that the bit is spinning at. (See also RPM.)

Fillet – Radius added to a square corner or edge. On corners, it can be used to round the 90° angle to the radius you select. But because router bits cannot cut square inside corners, many CAD/CAM programs can automatically add a "reverse fillet" opening and extra radius into the corner to cut beyond the radius of the bit, removing the rounded corner.

Flat depth – When carving with a V-bit, the depth of cut is variable and is defined by the bit meeting both sides of the vector being cut. The "flat depth" command allows for limiting the depth of the carving, preventing it from cutting deeper than desired.

G-code – A text file created from the CAD/CAM software that actually runs the CNC machine. This is a universal set of instructions that can be adapted to nearly any model of machine. Common commands all begin with the letter "G," which is where the name comes from. (See also post processor.)

Going home – Typically, this is a button command on a controller that automatically returns the machine head to the current zero or start point. This may or may not be the machine's mechanical origin point.

Hard home – Most CNC machines have a mechanical origin point that cannot be changed where all operations are computed from; I call this the hard home. It differs from other home positions because it cannot be easily changed, if at all. (See also home.)

Head – Component that holds the router or spindle to which the tool is attached. This section can move in the X-, Y-, and Z-axis.

Home – Refers to the mechanical origin of the machine itself, preset origin points that can be selected, or a predetermined start point for a particular program. (See also hard home, saved home, or temporary home.)

Homing – An automated process the machine may need to go through to physically find the mechanical origin point upon system startup.

Interpolate – Milling a hole to diameter using a bit that is smaller than the hole. The cut starts in the center and the bit curves outward to the final diameter. This improves the bit's life by reducing overheating and allows for easily adjusting the size of the hole being milled.

Jog – Manual movements of the head. This can be done by pressing a button and visually moving the head, or by entering a set distance for the head to move on command.

Lithophane – Originally porcelain art in which shallow figures are made visible by light shining through. In CNCs, this can be accomplished by carving photo-negative images in acrylic, showing the image when backlit.

Max carving depth – V carving toolpaths can cut quite deep if vectors are far apart. An overall depth can be set to limit the depth of the cut.

Onion skinning – Technique for holding workpieces that leaves a very thin (1/32"/1mm) layer of material behind by intentionally cutting too shallow. Tends to leave a cleaner edge on the part. (See also tabs.)

Post processor – A translation program that takes all of the data within the CAD/CAM program and rewrites it into a G-code file that is tailored to a specific type of CNC machine. This allows any CAD/CAM software to create G-codes that will properly control a wide variety of different CNC machines just by selecting the correct post processor.

Raster – Art files that are made up of pixels such as bitmaps and JPEGs. Rasters cannot be read nor used in programming CNC machines, they must be converted to vectors. Programs exist that can convert rasters to vectors, and most modern CAD/CAM software packages have some level of raster-to-vector conversion. (See also vectors.)

Relief groove – A cut made to remove much of the waste material from the path of an undercut bit. Relief grooves make for cleaner cuts and less overheating when forming an undercut groove. (See also undercut bit.)

Router – The motors from handheld routers often used to power the tools in an entry-level CNC machine as opposed to a dedicated spindle. (See also spindle.)

RPM – Revolutions per minute; a measurement of how rapidly the router bit is spinning as it cuts. (See also spindle speed.)

Saved home – Home positions that remain in the memory—even through system startup—until the operator resets them. (See also home.)

Scarf joint – Originally a shipbuilding term, a scarf joint joins two planks end-to-end using a long taper rather than a butt joint. With inlays, angling the edges of both the pocket and the inlay hides minor variations in the perimeter.

Secondary processes – Manufacturing steps that happen after milling on the CNC, including cutting tabs to remove parts from the web or adding threaded inserts into holes machined on the CNC.

Shank – Body of a router bit or end mill that is clamped into a collet when held in a router or spindle motor.

Single line fonts – In CNC programming, single line fonts are typically made from vector lines rather than the outlines from which TrueType lines are made. (See also TrueType fonts.)

Spindle – A motor that is purpose-built for use on a CNC machine. It has a collet to hold bits and serves the same function as when a handheld router motor is used. (See also router.)

Spindle speed – Essentially the same as the RPM setting on a handheld router, this is the rotational speed of the motor turning the router bit. (See also RPM.)

Spoil board – A secondary surface added to the machine bed of a CNC. The primary purpose is to allow router bits to cut through stock without damaging the bed of the CNC, hence the name; the board is there to be "spoiled" instead of the machine. The spoil board can also be used for work holding and be surfaced to provide an accurate Z-axis reference. (See also table mill.)

Start point – Also called the "datum," it is the base location from which all the features of a drawing or locations in a program are calculated. It is typically zero in the X-, Y-, and Z-axis. (See also datum.)

Tabs – Small, uncut sections left around the perimeter of a part to hold it in place but allow it to be easily separated from the surrounding web. (See also onion skinning.)

Table mill/surfacing – The process of using the CNC to mill a spoil board or other fixture to create an accurate Z-axis reference surface. (See also spoil board.)

Temporary home – A home position that can be set manually at any time. It is not saved if the system is shut down or a different home is selected. (See also home.)

Tool – Another word for the router bit used in the CNC for cutting operations; also known as "tooling." (See also bit.)

Toolpath – The set of instructions that control how a cut is made: the tool to use, depth of cut, feed rate, tool offset, and similar information. Toolpaths are divided into a number of categories that determine which parameters can be set and how. Common toolpaths include:

> **Profile** toolpaths are typically used for cutting parts out.
>
> **Pocket** toolpaths cut recesses in the stock where all the selected area is cut to the same depth.
>
> **VCarve** toolpaths incise letters and graphics using a V-bit. These are highly automated to mimic hand carving.
>
> **Drilling** toolpaths automate the creation of holes where the proper vectors are or in patterns that can be selected.

TrueType font – Letters made up as outlines rather than single lines. TrueType fonts work very well for V-carving. (See also single line fonts.)

Undercut bit – Slot-cutting bits that cut a larger interior opening than at the surface. (See also relief groove.)

Vector – Lines drawn between known points in a CAD drawing. Because all of the movement commands within a G-code program need specific X- and Y-coordinates, CAD/CAM programs require vectors for the drawings rather than image files, such as bitmaps or JPEGs. (See also rasters.)

Web – When parts are cut out of a larger blank, the leftover material from the borders and between the parts is known as the web because it holds everything together.

Work holding – How stock is clamped to the machine during milling. It can be as simple as double-faced tape or as complex as vacuum hold-down fixtures.

Working envelope – The area that the machine router bit can actually reach during operation. This is always smaller than the machine bed size.

Zero point – This term usually refers to the start point or origin of a program. (See also start point.)

SOURCES

CNC Machines

Axiom Precision: Manufacturers of the Axiom line of CNC machines. *https://www.axiomprecision.com*

Camaster: Manufacturers of industrial level CNC machines that offers a benchtop version as well. *https://www.camaster.com/cnc-routers/stinger-series*

Carbide 3D: Manufacturers of the Shapeoko and Nomad desktop CNCs. *https://carbide3d.com*

Carvewright: One of the very first benchtop CNC machines available, the Carvewright looks and works differently than most CNCs. *https://www.carvewright.com*

Gatton CNC: Kits, parts and instruction for building your own CNC machine. Building allows for a custom machine with a lower up front cost. *http://www.davegatton.com*

Inventables: Manufacturers of the X-Carve machines, they also have a free user friendly carving program to program their machines if carving is the main goal. *https://www.inventables.com*

Laguna Tools: Manufacturer of woodworking tools and equipment. They offer their IQ series of benchtop CNC machines. *https://lagunatools.com/cnc/iq-series*

Maslow: Not a manufacturer per se, Maslow is a unique open source, vertical CNC system that allows a larger format in a much smaller footprint. For those who like to tinker, this is worth a look. *https://www.maslowcnc.com*

Next Wave Automation: Manufacturers of the CNC Shark series of CNC machines. *https://www.nextwavecnc.com*

Powermatic: CNC machines made for Powermatic. *www.powermatic.com/us/en/c/cnc/cat_pwr_powermatic_cnc/?reset=true*

Shaper: Maker of the Origin handheld CNC. *https://www.shapertools.com*

Shopbot Tools: Manufacturers of industrial CNC machines. The Desktop series of machines offers a wide range of sizes and options in this category. *https://www.shopbottools.com/products/desktop*

Shop Saber: Industrial CNC manufacturer, their Shop Saber 23 is a benchtop model. *https://www.shopsabre.com/cnc/router/shopsabre-23*

Router Bits and Tooling

Amana: Manufacturer of router bits for use in many specific categories. *https://www.amanatool.com/products/cnc-router-bits.html*

CMT: Manufacturer of router bits for general use as well as a line of CNC bits. *https://www.cmtorangetools.com/na-en/router-cutters-chucks-for-cnc*

Freud: Manufactures a wide variety of router bits available through many retail outlets. *https://www.freudtools.com/explore/router-cnc*

Infinity Tools: Manufacturers of router bits for CNCs and handheld routers as well as accessories for general woodworking. *https://www.infinitytools.com*

Precise Bits: An online source for many hard to find engraving and carving router bits. *https://www.precisebits.com*

Onsrud: Industrial bit manufacturer with many specialized categories available. *https://www.cronsrud.com/store/router-bits.html*

Vortex: Manufacturer of industrial grade router bits including custom tools. *https://www.vortextool.com*

Whiteside: Manufacturer of router bits for hand use and CNC. Available through many woodworking retailers. *https://www.whitesiderouterbits.com*

Accessories

Rockler: Online and store-based woodworking retailer that carries CNC machines, router bits and many accessories for CNC machines. *https://www.rockler.com*

Woodcraft: Woodworking retailer offering CNC machines, bits and accessories online and in store. *https://www.woodcraft.com*

Software

Autodesk: Makers of AutoCad, Autodesk offers a huge range of professional level CAD and CAD/CAM software packages. ArtCam comes in different levels and is well suited to the small shop CNC user. *https://www.autodesk.com/products/artcam/compare/compare-products*

BobCAD: Long a favorite of those building their own CNC machines, BobCAD is still powerful enough to meet all the needs of home users as well as professional programmers. *https://bobcad.com*

Vectric: CAD/CAM software for almost any CNC brand or model, their VCarvePro is included with many hobby and professional CNC models. Vectric offers several program choices to fit different needs and budgets. *https://www.vectric.com*

INDEX

Note: Page numbers in *italics* indicate projects.